MAB Technical Notes 6

Titles in this series:

1. *The Sahel: ecological approaches to land use*

2. *Mediterranean forests and maquis: ecology, conservation and management*

3. *Human population problems in the biosphere: some research strategies and designs*

4. *Dynamic changes in terrestrial ecosystems: patterns of change, techniques for study and applications to management*

5. *Guidelines for field studies in environmental perception*

6. *Development of arid and semi-arid lands: obstacles and prospects*

Development of arid
and semi-arid lands:
obstacles and prospects

unesco

Launched by Unesco in 1970, the intergovernmental Programme on Man and the Biosphere (MAB) aims to develop within the natural and social sciences a basis for the rational use and conservation of the resources of the biosphere and for the improvement of the relationship between man and the environment. To achieve these objectives, the MAB Programme has adopted an integrated ecological approach for its research and training activities, centred around fourteen major international themes and designed for the solution of concrete management problems in the different types of ecosystems.

Published in 1977 by the United Nations
Educational, Scientific and cultural Organization
7, Place de Fontenoy, 75700 Paris
Printed by Union Typographique,
Villeneuve-Saint-Georges
ISBN : 92-3-101484-6
Développement des Régions Arides et
Semi-arides : Obstacles et Perspectives .
92-3-201484-X
© Unesco 1977
Printed in France

Preface

Approximately half the countries of the world are affected by problems of aridity. The diversity of physical, biological, socio-economic and political situations in these countries gives rise to a great variety of problems. Nevertheless, one problem is common to all arid zones: the fragility of the balance of arid ecosystems and the accompanying potential threat of desertification provoked in most cases by human intervention in these ecosystems. In practical terms, desertification, which is characterized by the spread of desert conditions beyond desert margins or by the intensification of desert conditions within arid regions, is accompanied by diminished productivity. In human terms, desertification may be seen as a lowered carrying capacity for livestock, diminishing crop yields, a progressive reduction in real income or in social well-being, and thus a reduction in the number of people supported in an arid region.

For over two decades, the international community has been actively concerned with problems of arid and semi-arid zones. From 1951 to 1962, a world-wide programme of arid zone studies was carried out by Unesco to promote and stimulate research in various scientific disciplines bearing upon the problems of arid regions. This programme resulted in the publication of some 30 volumes of *Arid Zone Research*, covering hydrology, soil salinity, plant ecology, human and animal ecology, climatology, energy resources, etc., in the development of major arid zone research and training centres and in the setting up of some 200 research units in 40 countries. In addition, during the past ten years, Unesco has been involved in the preparation of several thematic maps presenting up-to-date scientific knowledge on the natural resources of arid areas.

Current Unesco actions most directly related to the problems of arid and semi-arid zones are the Man and the Biosphere Programme (MAB) and the International Hydrological Programme (IHP). Two of the 14 major research themes under the MAB Programme are directly concerned with these problems. Project 3 deals with the impact of human activities and land use practices on grazing lands, including those in arid and semi-arid areas, while Project 4 concerns the impact of human activities on the dynamics of arid and semi-arid ecosystems, with particular attention to the effects of irrigation.

Recently, widespread drought and famine in the Sahelian zone of West Africa and in Eastern Africa have drawn world-wide attention to the problem of the management of arid and semi-arid areas. Two resolutions adopted by the United Nations in 1974 requested specific action with respect to this problem.

General Assembly Resolution 3337 (XXIX) on international co-operation to combat desertification assigned high priority to developing concerted international action to combat desertification, and requested the Secretary-General of the United Nations to convene a UN Conference on Desertification. The United Nations Environment Programme (UNEP) was given the responsibility of executing this resolution and is convening the Conference in Nairobi in August/September 1977.

In August 1974, the Economic and Social Council (ECOSOC), in Resolution 1898 (LVII), requested the Secretary-General to convene an *ad hoc* interagency task force on arid areas to identify problems encountered by developing countries which had not yet been overcome by science and technology. The task force was also asked to identify obstacles to the application of available technology, including social, economic, institutional and other obstacles, and to prepare an inventory of current research and development actions and programmes, with a view to preparing a world programme of development research and application of science and technology to solve the special problems of arid areas.

Subsequently, an Interagency Group on Arid Zones met at Geneva in October 1974 and decided that, in view of its experience in the field, the United Nations Educational, Scientific and Cultural Organization (Unesco) should assume responsibility for undertaking the study and for convening the task force in the form of a workshop, before mid-March 1975. The group also considered that the study should discuss the state of knowledge of arid zone problems and major trends in research and development; obstacles to the application of knowledge to the development of arid zones; and gaps and future needs in research and technology. It also indicated that emphasis should be laid on the need for integrated and interdisciplinary approaches to the subject.

Paul Pélissier, Professor of Tropical Geography at the University of Nanterre (Paris), was commissioned by Unesco to prepare a draft report in consultation with specialists from the Food and Agriculture Organization of the United Nations (FAO), the World Health Organization (WHO) and the World Meteorological Organization (WMO).

The workshop for the final elaboration of the report was held at Unesco headquarters in Paris from 30 January to 2 February 1975. Gilbert White, Director of the Institute of Behavioural Sciences at the University of Colorado, Boulder, Colorado (USA) chaired the meeting. The following specialists participated: Messrs. M. Batisse (Unesco); F. di Castri (Unesco); J.P. Charvet (France); J. Defay (Belgium); Miss G. Dehoux (Belgium); Messrs. B. Déom (WHO); J. Dresch (France); P. Dubresson (France); H. Einhaus (UN Office of Science and Technology); A.T. Grove (United Kingdom); R.N. Kaul (India); G. Lambert (WHO); H. Le Houérou (FAO); J. Lemoyne de Forges (France); J. Maunder (WMO); P. Pélissier (France); A. Sasson (Unesco); N.L. Veranneman (WMO). Messrs. M. Baumer (UNEP), R.G. Fontaine (FAO), M. Kassas (Egypt), V. Kovda (USSR) and R.A. Perry (Australia) were unable to attend the meeting, but submitted written comments on the draft report.

The report of this meeting was issued as document E/C:8 WG 1/3 of ECOSOC and considered by the intergovernmental working group of the Committee on Science and Technology for Development at its first session in April 1975. Because of a shortage of time, it was agreed that the report should concentrate on the examination of obstacles to development in arid and semi-arid zones, with the inventory of research and action programmes postponed to a later date. The working group welcomed the report and recommended that after revision in the light of its own comments and of those received from appropriate agencies or individuals, it be published by Unesco for broad dissemination. The document was thus amended to take into consideration the comments of the working group. Furthermore, certain parts of the report, namely those dealing with the physical characteristics of arid and semi-arid zones, irrigated agriculture, tourism, industrial and urban development, and transfer of technology have been dealt with in greater depth.

The present MAB Technical Note is thus based on this revised and expanded document. It attempts to highlight the main physical, sociological, institutional and biological features of arid and semi-arid lands, to identify the principal obstacles to the development of these zones and to propose a number of practical solutions as well as theoretical approaches and topics for research. As development of arid and semi-arid regions is associated with social and institutional changes and the application of new technology, it is anticipated that the publication of this Technical Note will help bring to the attention of those concerned with natural resources research and development of arid zones, as well as administrators and university students, the main problems associated with the management of arid and semi-arid lands and the guidelines for development within the particular physical and human environment of these areas.

Unesco wishes to thank all those who participated in the work of the Interagency Task Force as well as those who collaborated with Unesco in the preparation of this Technical Note.

Contents

I. Introduction. 9
    Delimitation of arid and semi-arid regions 9
    Low and variable precipitation 9
    Main factors of diversity. 11

II. Assessment of obstacles to development. 13
    Difficulties of animal husbandry 13
    Obstacles to the development of agriculture. 17
    Obstacles to urban, industrial and tourist development . . . 22

III. Development proposals 25
    Proposals for the arid zone. 25
    Proposals for semi-arid zones. 26
    Non-zonal features: industrialization and urbanization . . . 34

IV. Conclusion: limits and possibilities. 37
    Technology transfer. 38
    Necessary research 38
    Bibliography. 40

I. INTRODUCTION

The territory of half the nations in the world is located partly or wholly within the arid and semi-arid zones. Scarcity of rainfall, as well as its annual and interannual irregularity, are characteristic of these lands, which account for a third of the earth's surface and 15 per cent of its population (Batisse 1969; Sasson 1970).

Since 1950, at the initiative of Unesco, research on the problems of the arid zones has given rise to international collaboration. Among developing countries in the arid and semi-arid zones, the rate of economic growth has been rapid for oil-rich nations with small populations, but has otherwise been slow. In some regions, exceptional droughts have increasingly degraded essential resources and have emphasized acute socio-economic problems. It is important and urgent to determine the obstacles to development in these areas which stem from lack of technical and scientific knowledge, and those which result from a failure to apply it.

DELIMITATION OF ARID AND SEMI-ARID REGIONS

The transition from semi-arid to subhumid regions is generally gradual, except where it coincides with a distinctive geographical feature, such as a mountain range. It is thus best defined as a band of variable width, inside which very large oscillations occur. In the Sahel, for example, there may be a difference of several hundred kilometers between a 1972 isohyet and the same isohyet calculated as a 30-year mean. Included here are those regions which receive less than 600 mm of precipitation under summer tropical rainfall regimes and those which receive less than 400 mm under winter rainfall regimes at Mediterranean latitudes. Arid and semi-arid continental areas located at higher latitudes (in the "temperate" zone) are also included. The area thus delimited is marked by highly variable rainfall.

The transition from semi-arid to arid is mentioned here since solutions are proposed later in a zonal context. Regions where rainfed cultivation remains possible are considered semi-arid even though, because of climatic uncertainty, it is usually high-risk cultivation that is practised. As this definition is rather vague, the transition between arid and semi-arid can be schematically located at the 250 mm isohyet. In practical terms, however, as climatic averages do not exist in a real life situation, it is more accurate to speak of the probability of certain phenomena, such as, for example, three consecutive rainy seasons with rainfall inferior to the long-term mean. The extreme variability of precipitation in time and space is in fact an essential characteristic of these regions.

LOW AND VARIABLE PRECIPITATION

Spatial distribution

In terms of the spatial distribution of rainfall, growing aridity is detected through the study of isohyets. These isohyets are means calculated from statistical series of varying length, depending on the stations. But in a single rainy season, there are very marked irregularities in space. For example:
- between the stations of Bakel and Dakar (Senegal), which are both located on the 500 mm isohyet, a difference of 270 mm was recorded in 1972;
- over shorter distance, Atar and Akjoujt (Mauritania), which are about 150 km apart and in a comparable climatic area, recorded a 160 mm difference in 1970, which is more than the annual mean of either of these

stations (Académie des Sciences d'Outre-Mer 1975).

Variations in time

Rainfall varies over time, both in its seasonal distribution and its distribution between years. A map of global interannual variability of precipitation such as Petterssen's (Petterssen 1941) clearly shows the arid and semi-arid regions. Variability of rainfall generally increases with aridity. Thus relative variability of precipitation (the ratio of the mean deviation to the annual mean, multiplied by 100) is much higher in the central Sahara (79 per cent at Djanet, 92 per cent at Adrar) or on the Red Sea coast (127 per cent at Qseir) than it is on the northern edge of the Sahara (40 per cent at Biskra, 30 per cent at Colomb Bechar). Absolute values are even more indicative, especially in extreme cases. At Swakopmund on the Namib coast, over a period of 25 years the records show one year with 148 mm of rainfall and another with only 1 mm. Finally, the uniqueness of a given station in relation even to neighbouring stations can be shown during "surplus" years as well as in dry years. In 1957, during a particularly wet period in Niger, only 64 mm fell at Agades, where the annual mean is 164 mm. This is less than fell during the drought years 1971 and 1972, when 94 mm and 74 mm were recorded respectively. On the other hand, in 1958 the rainfall at Agades was 297 mm.

The distribution of rainfall during the rainy season, which has a direct influence on dryland crops, is an additional factor of variation. At Aritunga, for example, in the middle of the Australian desert below the Tropic of Capricorn, the average rainfall for the month of March is 52 mm; however, March 1910 had 340 mm and March 1911 only 5 mm. Although the annual totals for 2 years may be very similar (600 mm in 1956 and 1957 at Zinder, Niger), the structure of the rainy season is often very different. In 1956 rain did not start until the month of June (33 mm) and 535 mm were recorded in July and August; in 1957, on the other hand, heavy rain was recorded as early as May (78 mm) and June (101 mm) but only 388 mm in July and August, which on the whole is a much better distribution in time.

Intensity of evaporation

The effects of irregularity and variability of precipitation on water resources and vegetation are further aggravated by the intensity of evaporation. A recent study of the drought in the Sahel, involving 78 experimental basins (Académie des Sciences d'Outre-Mer 1975), shows that evaporation varies approximately inversely with precipitation in this region: as rainfall decreases, evaporation increases. Therefore aridity increases much faster than the isohyets indicate. The consequences are all the more important because the largely bare and overheated soil is ill-suited to store early rainfall. This is why erratic rainfall at the beginning of the rainy season may seriously endanger dryland crops. A precise measure of aridity still remains very difficult because of the great variety of climatic data which must be taken into account. Many different indices have been suggested and used; in the great majority of cases the limits they indicate are very similar, especially on a small scale.

Main factors creating aridity

It is important to bear in mind the main factors which create aridity on a worldwide scale. The great arid zones most often result from a combination of several of these factors, as is shown by the distribution of the arid belts (for example, the belt from the Sahara through the desert of Arabia, Iraq, Iran and Turkestan to the Gobi). The high pressure belts located at tropical latitudes are primarily responsible for aridity; but there are other elements to be considered, including:
- continentality: the further one penetrates into compact continental masses, the smaller the chance of abundant precipitation (deserts of Central Asia);
- the existence of mountain ranges which hold back rainbearing winds and in addition cause foehn winds which increase evaporation on the "leeward" slopes (semi-arid zones located in the shadow of the Andes and the Rocky Mountains);
- the effects of cold oceanic currents along some coasts (the coastal deserts of Peru, Chile and southern Africa).

The great variety of situations in the arid and semi-arid world is determined not only by climate, but also by topography, hydrography and soil conditions, different types of social organization, the history and types of land use, the nature and availability of the mineral resources, etc. Possible classifications must take into account the great number of possible combinations of the numerous natural parameters and the diversity of human situations resulting from socio-cultural settings, political divisions and the historical conditions of development.

MAIN FACTORS OF DIVERSITY

Climatic conditions

Diverse climatic conditions give rise to very different situations. Beyond a simple classification into arid and semi-arid regions, it is important to consider the duration of the sparse rainfall, and the duration and timing of the dry season. It is obvious, for example, that the problems of irrigation, dryland farming and pastoral movements will be resolved differently on the northern fringes of the Sahara where the rain falls in winter and on the southern fringes where the rain falls in summer. In the case of rainfed cultivation, dryland farming techniques are well suited to zones with winter rainfall, since evaporation is then relatively low and part of the water can be stored in the ground and used by the plants the following year. On the other hand, these techniques are not very effective in zones of summer rainfall because evaporation is very high at this time of year. The climatic originality of a given semi-arid zone thus appears to be largely a function of the general characteristics of the adjacent subhumid zone, and an important factor in creating diversity.

Moreover, the precipitation recorded by rain gauges is not always the only source of water; one of the original features of arid and semi-arid coastal or subcoastal regions is the presence of hidden precipitation (such as dew), resulting from a greater humidity of the atmosphere (Chile, Mauritania, Morocco, Tunisia, etc.).

Finally, temperatures vary widely from one arid or semi-arid region to another: hot deserts lying on the tropics (the Sahara and deserts of Arabia) are often contrasted with temperate zone deserts (such as those of Turkestan), characterized by cold winters and by very great annual temperature variation because of their continental location. In general, crossing the tropics towards higher latitudes, the annual range of temperatures becomes greater than the diurnal range. In hot deserts, the annual range is between 15 and 25 °C (24.4 °C at Timimoun in the Algerian Sahara); it is markedly higher in the continental deserts of the temperate zone (37.7 °C at Turgai in Central Asia) where the range of extremes is exceptional (sometimes over 90 °C). As a consequence, the harshness of winter temperatures in the steppes of Central Asia creates conditions altogether different from the relatively mild ones in the steppes on the edge of the Mediterranean world, where the constraints on man are not as great.

Likewise, the exceptional heat of some deserts (in absolute value as well as duration) imposes different types of adaptation on animal life and exercises physiological and neurophysiological constraints on man which limit his activity.

Finally, as the intensity of evaporation is a function of the temperature, the amount of water actually available for plant growth varies considerably for the same volume of precipitation, depending on the latitude and continentality of the region considered.

Allogenic rivers and water courses

The presence of allogenic rivers, rising in non-arid regions (the Colorado in the United States, the Amu Daria and Syr Daria in Soviet Central Asia, the Indus in Pakistan, the Nile in Egypt, the wadis which flow down from the Atlas in North Africa, etc.) may radically modify the development of a region. These rivers have often made it possible to considerably extend areas under irrigation, but in other cases they are still very incompletely used (the Niger, the Logone-Chari, the Senegal, etc.) or could be used differently.

The flow characteristics of these rivers introduce considerable variation. For example, the flow of the Nile, depending mainly on tropical rainfall, is very different from that of the mountain rivers of higher latitudes, which are liable to snow retention, combined in places with Karst retention. Further, use of surface and underground water varies according to topography and geology; thus the Sahara and Arabia contrast with Central Asia and the Andean countries.

In the case of groundwater, essential to agriculture and to human settlement, a distinction must be made between regions where the underlying geological base is of generally hard and impermeable crystalline rocks and those where it is of sedimentary rocks of varied composition. In ancient crystalline terrains ("shields"), groundwater is of local importance only and its reserves are limited; it is also sometimes difficult to assess its size and pressure precisely. In this case, shallow underground water flow (inferoflux) in alluvium produced by mechanical disintegration of the crystalline base is important; however, even if the wadi in which it takes place is relatively well-fed for part of the year, it remains subject to climatic variations precisely because of its limited size.

In sedimentary terrains, on the contrary, groundwater tables may be large and may be found at several different levels (the water tables of the intercalary continental zone and of the terminal continental zone in the Sahara, separated by terrain not suited to water storage).

In order to use these resources, it is essential to distinguish between renewable water tables, constantly recharged although often irregularly, and fossil water tables (not now being recharged). The advantages of sedimentary basins over crystalline bases and shields are obvious, and the geological structure of arid and semi-arid regions is thus an important factor in determining water availability.

Soil conditions

Soil conditions vary greatly according to the distribution of the parent rocks, climatic conditions and topography. In many instances there is no real soil and the original substratum is apparent. Where soil does exist, sandy soils seem best suited for storing water from the rare and irregular rainfall. In irrigated zones, silty soils are in some respects better than sandy soils, but they present severe risks of soil degradation, especially where natural drainage is poor (the Nile delta, plains of the lower Euphrates, etc.).

Accumulation of salts ($NaCl$, CO_3Na_2, SO_4Na_2, SO_4Ca, etc.) is generally important because of high evaporation, and closed basin drainage in many regions. This concentration of salts is at its worst at the centre of depressions where, because of climate and topography, rivers cannot flow to the sea. Halophilic plants alone are able to develop in these conditions (*playas* of North America, *takyrs* of Central Asia, *sebkhas* of North Africa).

Finally, all soils of the arid and semi-arid zones remain, to varying degrees, vulnerable to erosion, whence the overriding importance of all operations of soil fixation, conservation and restoration. These operations are even more urgent since in some cases there are fossil soils, created under different climatic conditions, which are thus unable to reconstitute themselves if they are destroyed.

Continental position

The continental situation of many arid and semi-arid zones influences development possibilities insofar as modern development implies contact with the outside world. Thus, coastal regions have an advantage over regions located thousands of miles from the sea. The ease and frequency of contacts with technologically advanced rich countries, or even with the vital centres of a nation, are often decisive. This is illustrated by the contrast between the Chad Basin and northern Mexico.

Investments

The position occupied by the arid and semi-arid regions within the countries concerned (marginal, important, essential, or even vital) is a factor determining the attention governments pay to them, and, in particular, the public or private investments available for their development. Such investments are rarely attracted by areas where the immediate return is much less certain than elsewhere.

Population density

Population density is an essential element of differentiation. In Australia, for example, the development of semi-arid regions is hindered by shortage of labour. In other regions, the poverty of the inhabitants is due to their great number relative to environmental possibilities in a given technical, economic and social context. Large human or animal populations can provoke dangerous overexploitation and cause serious damage to the environment. The maps of livestock and population distribution established by the Société pour l'Etude du développement économique et social (DGRST-SEDES 1976) for the Sahel emphasize some important contrasts and enable the most endangered zones to be identified.

A number of other factors can, however, create very different development problems in areas with similar population densities and comparable natural conditions. In different cultural settings (North American, South American, Muslim, Australian), the environment and its problems can be perceived in different ways. Proposals for change should be adapted accordingly.

The level of economic, social and technical development is another factor which can widen or reduce the range of possible choices, and to a certain extent affect the distribution of population in rural or urban areas.

Lastly, since social and psychological problems of adaptation to new ways of life are greater than technical ones, the approach to development will differ according to the proportion of nomads, semi-nomads and sedentary peoples within a given region.

Under these conditions, rather than encourage large concentrations of population around irrigable areas, it would seem preferable to plan more varied and comprehensive development measures. Furthermore, given that in some countries the technical context has little chance of improvement, even in the medium term and that growing population pressure will certainly aggravate the consequences of unpredictable climatic fluctuations, a fundamental question for the future of arid and semi-arid regions can be raised: where choice is possible, should surplus population be maintained in its original environment, or should it be, on the contrary, relocated in more favourable areas?

Mineral and energy resources

The availability of mineral and energy resources, as well as the conditions of their use, introduce new divisions. Access and extraction are important, but price changes on the world market may make profitable the mining of previously unprofitable deposits, or greatly increase the value to states of mineral fuels or raw materials. Possibilities for local processing obviously have a direct bearing on economic and social development. The intermediate stages of the processing of industrial products are generally poorly represented in the arid and semi-arid zones.
In some regions mining industries, or basic processing industries (concentration of an ore for example), predominate, while in some oases there are highly technical light industries (such as electronics), as in the United States. Finally, the recent development of tourism has definitely been advantageous for some areas. This development depends on a number of factors (presence of beaches or archaeological sites, ease of communications with the countries which "supply" tourists) which create a great variety of situations.

This great diversity renders extremely difficult the elaboration of development models which ignore regional characteristics. However, given common constraints, linked to the irregular supply of water and the consequent insecurity of human use, the most severe problems arise not in arid, but in semi-arid zones. There are specific solutions for arid zones, based on division into specialized land use units. In semi-arid zones, on the other hand, a choice must be made among various methods of development (agriculture or animal husbandry), a choice even more difficult with increasing population densities.

In the first part of this Technical Note the main obstacles to development are discussed:
- problems of animal husbandry;
- obstacles to the development of agriculture;
- obstacles to urban and industrial development, and to the spread of tourism.

The fact that these problems are presented separately in no way implies a search for purely sectoral solutions. In the second part, some practical solutions and research topics are proposed.

The term "development" is not used to mean specific achievement or a particular technical feat, but rather any measure which can be diffused within a population to improve its general standard of living. Attention is focussed on problems specific to arid lands in both developing and developed countries.

II. ASSESSMENT OF OBSTACLES TO DEVELOPMENT

DIFFICULTIES OF ANIMAL HUSBANDRY

Pastoralism plays a vital role because animal husbandry is capable of using nearly all the land, and is often an essential sector of the economy. It also supports civilizations based on different forms of nomadism. It can be estimated that the arid and semi-arid zones as a whole contain over half the world's stock of cattle, more than a third of its sheep and two-thirds of its goats. Nevertheless, livestock productivity in non-industrialized countries is only 10 to 20 per cent of that obtained in modern animal husbandry. Vital, but of low productivity, animal husbandry in arid and semi-arid countries is thus extremely vulnerable. The hecatombs resulting from the recent drought in Ethiopia have provided a dramatic reminder of this vulnerability.

What are, therefore, the main difficulties involved in the rationalization and, in some cases, the preservation of this age-old and still vital activity? The most intractable difficulties are caused by the natural environment itself. Climate imposes severe limits on resources available to water and feed livestock and causes profound insecurity. The recent years of drought in the African Sahel are a reminder of the permanent difficulties that animal husbandry has to face and which justify its techniques - particularly its mobility.

Characteristics of rangelands

The seasonal and often sporadic nature of the useful rainfall, the length and severity of the dry season, the intensity of evaporation and the great variability of rainfall from one year to another mean that water resources are uncertain and intermittent. Because of uneven spatial distribution of these resources, large areas of rangeland can be used only during the short rainy season when temporary pools abound. For the greater part of the year, the herds must withdraw to areas with a moister climate (either at different latitude or at different altitude) or must gather in allogenic valleys or in areas where water points and pasture are easily accessible. Climatic differences and the possibilities offered by regional contrasts resulting from the configuration of the land create a considerable diversity of pastoral systems and of nomadic movements.

Technical progress, the sinking of deep boreholes in particular, has increased the number of permanent water points. But this only improves watering of livestock which feed almost entirely on natural vegetation. Although there is considerable diversity of natural pastures, their carrying capacity for livestock is invariably low and falls rapidly as aridity increases. Overstocking is all the more serious since the plant cover is more fragile.

Secondly, drought reduces or destroys seasonal grass growth and encourages the spread of bushes, often halophytes, in sub-desert winter rainfall steppes (from the Maghreb to Iran), and thorn-bushes, particularly Acacias, which cover immense stretches of steppe from Mauritania to the Thar desert or, in the southern hemisphere, from the Argentinian Chaco to the Kalahari and south Madagascar. Although there is also some seasonal grass vegetation, the length of the dry season combined with high temperatures and considerable solar radiation rapidly transforms it into a discontinuous cover of little grazing value. The only regular appearance of dense stands of rapidly growing grasses is on loamy soils deposited by the flood waters of the inland deltas of the major allogenic rivers such as the Niger, the Logone-Chari, the Nile or the Zambezi.

The third factor common to rangelands in arid and semi-arid zones, which stems from the preceding one, is that the food which it provides for livestock varies considerably both in quality and quantity at different times of the year. Abundant and sometimes rich in the rainy season, it very quickly becomes inadequate and of low nutritive value during the long dry season when herds lose weight, become weak and are increasingly prone to disease.

An additional difficulty linked with the preceding problems results from the frequent discrepancy between water and pasture. This has been aggravated during recent decades by different policies for providing water; these have resulted in localized improvements in the watering of livestock, thus leading to abnormal concentrations of animals which have brought about, through overgrazing, the deterioration or destruction of pasture. This has frequently been the case around the many deep boreholes drilled since the last world war. In the regions affected by the recent exceptional droughts, the considerable and at times catastrophic losses of livestock were caused by starvation and only rarely by thirst.

Crisis of the traditional pastoral way of life

The response by pastoralists in arid and semi-arid regions to these difficulties has always been, and remains, geographical mobility combined with an adjustment of the density of the human and animal population and of the composition of herds to bring them into line with the carrying capacity and thus with the limited and precarious potential of the various environments. This mobility takes such varied forms that any classification would be arbitrary. In very arid regions, pastoral movements do not follow any kind of periodic pattern and are as irregular as the rain itself. On the other hand, in mediterranean and tropical marginal zones, the mobility of pastoralists and herds has a seasonal rhythm; movement, often over long distances, takes place between zones with complementary climates. This is the case, for example, of pastoralists from the Saharan piedmont who spend the summer in the Tell, or of the symmetrical movements of the Twareg of the southern Sahara. A third type of mobility is made possible where areas with greatly differing climatic conditions are found side by side at different altitudes. Such patterns of movement are found, for example, in the mountain regions of Iran and Afghanistan and in the Andean region and its arid fringes. It is only on the edge of the semi-arid zone, or where the relief or the presence of allogenic watercourses create exceptional ecological conditions, that sedentary animal husbandry can exist as part of relatively integrated systems of agricultural production.

But the delicate balances achieved or sought by traditional forms of pastoralism have been upset everywhere during recent decades. The general deterioration of rangelands bears witness to this phenomenon. The situation of animal husbandry and of those who practise it thus becomes a critical one, caused by man himself and characterized by so many of the obstacles which hinder the functioning and, *a fortiori*, the development of pastoral economies.

Difficulties of animal husbandry

The main causes of this crisis are overpopulation of people and animals in the countries in question, particularly in the semi-arid regions where crop production and animal husbandry become competitive rather than complementary forms of land use. These pressures are partly attributable to the demographic growth which characterizes the populations of almost all the countries of the Third World. Although all the data combine to show that population growth among groups which practise different forms of nomadism is noticeably lower than among settled farmers, the arid and semi-arid zones continue, as they have done throughout history, to produce a surplus population. Because of the poverty of the environment, this growth in population leads to overexploitation and emigration. Simultaneously, the size of herds has increased as a result of improved health conditions, the provision of numerous water points and the setting up of a modern infrastructure, one of the effects of which has been to facilitate the development of different forms of semi-nomadism, sometimes motorized (in the Libyan desert and Andean piedmonts).

At the same time, the semi-arid belt, where traditionally the surplus pastoral population was absorbed and settled, has witnessed an unprecedented growth in sedentary population. Furthermore, farmers have everywhere sought to extend their cultivated land, to grow not only subsistence crops, but also various cash crops. Lastly, the emancipation of certain social groups which were previously an integrated part of pastoral societies has been accompanied by their settlement. These new members of the sedentary population have become farmers who have, in turn, taken over land at the expense of the flocks and herds. Under the combined pressure of increased man-power availability and increased needs, and in some cases using modern equipment (tractors, for example), "pioneer fronts" have been opened up which have eaten into the traditional domain of the pastoralists from the Atlantic to Iran, and in Latin America, the rangelands have receded during the last three decades in the face of a disorderly expansion of new agricultural holdings.

This veritable race for the use of land has extended the rangeland areas towards the least hospitable zones. The African Sahel presents the most striking example of this trend, a trend encouraged by the relatively humid climatic phase which occurred between the two most recent drought periods - that which lasted from 1940 to 1944 and that which started in 1966. Thus, over a period of about 20 years, there has been a general shift towards the north (in other words, from semi-arid to arid zones) of all those peoples whose main occupation is pastoralism. In their wake, sedentary peoples have settled at latitudes which, for centuries, had been exclusively the province of a pastoral way of life. The case of the Fulani pastoralists of the Sahel is typical in this respect. The pressure of the farmers, leading to their regular invasion of pasture land, has reached such proportions that in one of the few countries (Niger) where public authorities have shown any concern for the pastoralists' interests, legislation to limit the clearing of ground for farming, had been introduced well before the start of the recent drought.

But, what meaning can there really be in a boundary line drawn through an area as shifting as the Sahel? For instance, it has been shown that, within the space of ten years, a third of the immense territory of Mauritania was either desert or pasture land attracting herds of livestock. The withdrawal of pastoralism towards the most arid regions which the farmers could not exploit, is largely responsible for the vulnerability of livestock breeding and for the disastrous effects of the recent drought in the Sahel.

The pastoralists, now isolated in the least accessible regions, are also finding that the pastures where they used to go in the dry season and where they could find water and forage, are now gradually being lost to them. For instance, cattle used to have access to the stubble in the fields as soon as the rainy season ended, but today the ground is taken up by cotton for several months of the dry season. Similarly, the banks of rivers and pools are increasingly invaded either by flood-retreat crops or by rice paddies, so that the free access of herds to water and the lush pastures of the flooded lands is being disputed more and more. It even happens that modern agricultural developments suddenly cut off the most valuable pastures which had always been the livestock breeders' last resort (e.g. in the Awash valley in Ethiopia).

All these events lead to conflict. Tensions and sometimes clashes occur in zones still reserved for nomads, where the original inhabitants are now in competition with other pastoralists pushed out of their pastures by the farmers. Many conflicts arise between nomadic pastoralists and sedentary cattle breeders who are trying to stop the movements of herds, or at least to force them out of those areas now taken up by crops.

The pastoralists feel the geographically marginal position to which they have been relegated all the more keenly in that it has been accompanied by a reversal of their political and social status, and this frequently seems to be an obstacle to the expansion of livestock raising.

Historically speaking, nomadic pastoralists have played a preeminent role, not only in the deserts, where they controlled all movements, and the oases, but also at the edge of deserts where they settled their surplus numbers and where their military superiority often gave them control over the sedentary peoples. In addition, the co-existence of pastoralism and rainfed cultivation, and control of the intrusion by pastoralists into lands belonging to sedentary farmers, was achieved with the greatest efficiency by the firm, centralized organization of large nomadic confederations, as in Iran, for example, in the Zagros and its border regions. This historical superiority, however, even in the States of the Arabian peninsula, has today disappeared. A much studied example of this development and its consequences is provided by the Twareg of the Sahel in Niger (Bernus and Bernus 1972).

An additional factor is that nomadic societies everywhere have fertility rates, and particularly population growth rates, noticeably lower than those of the farming communities near or among whom they live. The growth in their numbers is thus far slower than that of sedentary peoples. Furthermore, within a given cultural community, it seems that the more the various groups confine themselves exclusively to pastoral activities, the lower are their respective growth rates, and that the latter are higher the more the people are integrated in an agricultural economy.

Of all the factors which could explain this change, the most general and the most objective are no doubt the under-equipment and under-administration of the arid zones, the peripheral nature of their economy and, finally, the marginalisation of their population. The attitude of States towards nomads and, more generally speaking, towards stockbreeders, is still responsible in most cases for the lack of concern shown for the pastoral way of life. This attitude undoubtedly impedes the activity which is best adapted to the severity of the environment.

Mobility remains the most effective adaptation to the use of sparse resources which are sporadic and short-lived. Yet by virtue of its nature, aims and methods, the modern State, whatever its political options, is, owing to its very structure, opposed to this mobility.

The very few exceptions - and in fact there is only one worthy of the name - do not disprove this rule and are of a circumstantial, and therefore temporary, nature. This hostility of any modern government to the mobility of pastoralists is found under all regimes and the aim, sometimes explicitly stated as in Syria, is to do away with nomadism and its attendant social organization. Even governments stemming from the Bedouin civilization, as in Jordan and Saudi Arabia, are encouraging all their peoples to settle and do not oppose the encroachment by various forms of speculative agriculture on traditionally nomadic lands. The Afghan State is an exception to this, as it continues to support nomadic life, at least that of the peoples whose mobility contributes to the integration of the mountain areas of Central Afghanistan into the life of the nation.

Action against the nomadic use of land is supported by a whole arsenal of legislative and administrative measures. These measures are all designed to restrict, hinder and control the movements of men and herds. This action is also backed up by the provision of facilities adding to the attractions of sedentary life, whether agricultural or urban. But it is above all remarkable that most of the investments ostensibly destined for pastoralists have, with the consent and sometimes the support of governments, been diverted from their original objective to benefit not only the settlement of nomads but agriculture as well. Thus, deep boreholes have often attracted farmers in such large numbers that the wells have become inaccessible to the livestock herds. In the Ferlo region of Senegal, the drilling of wells has speeded up the eastward migration of the farmers and has contributed to the reduction of the land available for grazing. Here, as elsewhere, public authorities almost never take the side of the pastoralists, and the sinking of wells has opened up lands, hitherto exclusively devoted to pastoralism and to agricultural occupation and use.

In some areas, even the systems of taxation are detrimental to pastoralism. In Africa south of the Sahara, for example, livestock is subject to a tax per head whilst farmland is exempt from taxation. Pastoralists resent this distinction between the two sources of production as a discriminatory prejudice which helps to put them in a marginal position in the national community and penalizes their activities.

Economic problems

Finally, there are economic obstacles to the development of pastoralism. There is no activity which brings in less money or whose profits are more chancy. This situation is the result, first of all, of the isolation in which specialist pastoralists live and, consequently, the considerable distances which generally lie between them and the centres of consumption. It is also the result of trading conditions, since trade is usually in the hands of intermediaries who have well-organized purchasing and sales networks and who extract a larger profit from the livestock than the producers themselves. The producers are also penalized by the fact that they trade in live animals which can seldom be delivered directly to the butchers. Finally, the extreme irregularity of supply, linked with the producers' lack of organization, the seasonal nature of their visits to markets and of their cash requirements, is one of the main reasons why livestock raising does not yield good returns. Although the meat and milk markets are largely available, supply is sometimes overabundant but more often deficient, so that a country which is a potential exporter imports in order to supply its cities. Such disorganization in marketing has distressing effects on prices, disheartening the producers.

In this connexion, a commonly held fallacy should once again be denounced with few exceptions, the general attitude of pastoralists, particularly the more highly specialized among them, is not hostile to the commercialization of livestock. Their wish to accumulate as many animals as possible and to keep what is often thought to be an excessive number of old animals is principally a precaution against the hazards of their environment. If a profitable market were open to them and if elementary guarantees were provided, such as those from which many farmers benefit, they would trade all the more willingly. They are perfectly conscious of the need to improve their pastoral operations by lightening the pressure exerted by livestock on the rangeland.

It is the marginal position of the pastoralists and the peripheral nature of their production which seem to be the fundamental obstacles to the increase of production and to the improvement of the standard of living of the pastoralists themselves. In countries lacking a pastoral tradition, these difficulties have been overcome through a specialization of livestock raising linked to a rational marketing system at the national level. In this way, the Patagonian regions of Argentina and Chile have been integrated in their respective national economies.

OBSTACLES TO THE DEVELOPMENT OF AGRICULTURE

Difficulties of irrigation

Irrigation is too often proposed as a solution to the problems of arid and semi-arid areas without sufficient attention being given to the fact that agriculture in these conditions is costly and requires sound experience, as well as specific precautions. There are numerous examples of soils degraded and lost to production following ill-conceived or poorly implemented irrigation schemes. The most serious difficulties relate not to the detection and delivery of water, but rather to the secondary effects of irrigation, which compromise success in the long-run.

Several sets of obstacles must be distinguished.

Irregularities of water supply

The irregularity of water supply is even more limiting than the small supply of water. A recent study of the rural population in Morocco shows that a given quantity of water will support three times as many people if it is distributed regularly instead of in the form of floodwater (Noin 1971). One solution lies in the accumulation of water behind dams. However, since the seasonal irregularity of rainfall is complicated by marked inter-annual differences, dams - to be really efficient - must be able to contain at least a year's supply of water and, if possible, more than one; hence, the gigantic size of some projects. The reservoirs built up by dams raise at least two technical problems. First, there is the problem of silting, which may take place very rapidly when the drainage basin has not been previously managed against the effects of erosion. Second, there is the problem of intense evaporation from these reservoirs. Since the volume of water which evaporates is proportional to the surface area of the reservoir, the larger the water body the greater are the losses. Water is thus lost at the same time as salinity in the reservoir increases.

Management of water resources

Another difficulty is connected with the management of the water resources. On the average, 50 per cent of irrigation water is lost by infiltration and evaporation before it reaches cultivated plants, though the use of underground pipes and cemented channels helps to reduce these losses. The most catastrophic consequence of inadequate water management techniques is the creation or restocking of the underground water table which rapidly induces waterlogging of the soil. Every irrigation project, especially in low-lying plains, should be accompanied by hydrological studies, by surveillance, and, if need be, by systematic drainage, planned in advance and not improvised afterwards. It is essential to control the rise of groundwater. This implies a rise in project costs, but guarantees medium-term success.

The quality of irrigation water should also be controlled. If it is not satisfactory, it is necessary to leach the soil. This increases in the passage of salt-laden waters downstream and leads to special difficulties when these waters cross a national boundary.

Salinization

Salinization can lead to the alkalinization of soils, an even more serious problem in view of the major difficulty of reversing the process. Relatively heavy soils are more sensitive to salinization and alkalinization than are sandy soils. These heavy soils are most frequently found in low-lying plains, in the very areas where it is most difficult to install a drainage system and to stop the rise of groundwater.

Zones vulnerable to damage through saturation and salinization represent 50 per cent of the total irrigable area in Iraq, 23 per cent in Pakistan, 50 per cent in the valley of the Euphrates in Syria, and 30 per cent in Egypt. In Syria the area under cotton has increased rapidly under irrigation, but, in the absence of drainage, yields have fallen by a good third, due to increased salinity and saturation of the soils. In Pakistan it was estimated in 1963, when corrective measures began, that more than 40,000 ha of productive land were damaged each year by waterlogging and increased salinity. In Egypt, the transition to permanent irrigation, entailing markedly increased inputs of water into irrigated zones, has caused groundwater in many of the cultivated sectors of the delta to rise to a level that is critical for most crops.

Soils

Cultivation of the soils of arid and semi-arid zones is generally a very uncertain operation. These soils are rarely well-balanced, are low in organic matter, and have a marked structural instability and a high density. The fact that they remain moist under permanent irrigation aggravates the problem. Considerable care is required in working moist soils, and an error can cause damage which is difficult to repair. Strong drying-out of the soil would improve the situation, but this would entail temporary abandonment of cultivation and would prove difficult to implement if only for economic reasons. It should be stressed that, overall, the addition of water through irrigation has an unfavourable effect on soil structure. Finally, it should be noted that in certain cases the factor which limits production is not the lack of water but rather the nature of the soil to be irrigated, as in the dry, north-eastern parts of Brazil where the waters of the River Sao Francisco are largely unsuited for irrigation purposes.

Economic value

The question of the real economic value of solutions to land use based on irrigation is not an easy one to resolve. In many countries the most immediately profitable operations have already been carried out. New irrigation works should almost invariably be accompanied by the installation of an efficient system of drainage, and hence require additional investments. Generally speaking and taking into account the increased costs of necessary materials, it may be stated that the opening of new irrigated areas will be more and more costly. Moreover, it is often preferable to improve existing irrigated plots rather than to create new irrigated areas. Many countries have reached the point where the cultivation of food crops alone is not enough to provide the necessary surplus to make investments profitable. It should be noted in this respect that irrigated agriculture practised in modern installations involving large investments will increasingly become synonymous with scientifically-based commercial agriculture, in other words a form of agriculture which requires well-trained personnel.

However, the bias which the governments of the countries concerned (as well as the countries aiding them in various ways) seem to have in favour of large projects is often more connected with prestige at home and abroad than the result of detailed analysis of the economic effectiveness of the project. In a certain number of cases at least, modest and relatively less costly projects would have a greater impact on the economy and improvement of the lot of the people. However, one potentially significant consequence of hydraulic works is the development of fish farming. Its basic requirements are sufficient motivation and a minimum amount of training of a section of the population.

The combination of rice and fish farming, which has long been practised in some countries, could yield good results. This technique, which uses plots covered by 10 to 50 cm of water during the vegetative cycle of rice (an environment very similar to that found in the flood plains of certain allogenic rivers like the Niger), could produce considerable food surpluses. One of its main shortcomings, however, is the necessity for chemical treatment of plants. Little is known so far about the effects of herbicides and insecticides on fish (Worthington 1977).

Legal or social obstacles

Obstacles of a legal or social nature can hinder possible initiatives for irrigating new land areas: for example, land areas which are not propitious for even partial mechanisation, complex and rigid rights of water use and tenant farming under unjust contractual conditions for the tenant. In general, such situations are found in zones which have long been inhabited by man, but in a number of recently opened lands difficulties are met in circumventing traditional social structures.

Badly-organized irrigation may increase the numbers of malarial mosquitoes or the snails that are hosts to the schistosoma causing schistosomiasis. The multiplication of these vectors is particularly favoured by the presence of stagnant or slow-moving water colonized by aquatic plants. While it is incorrect to assert that irrigation itself increases the incidence of schistosomiasis or malaria, it is certain that poor maintenance (or bad design) of irrigation installations may provoke the formation of pools and the accumulation of useless water, together with the growth of plants which harbour the larvae of the *Anopheles* mosquito and snails that are hosts to the schistosoma. It should nevertheless be noted that the spread of these snails is linked to the chemical composition of the water. This explains the absence of these animals in some parts of the world, such as Peru and Chile. In other regions, however, some irrigated areas have had to be abandoned in the past because of the increase in diseases caused by poor maintenance of irrigation works.

In any case, it is certain that irrigated agriculture, which calls for great expenditure of human energy, requires healthy workers. Malaria or schistosomiasis may be brought into newly irrigated regions by workers coming from previously irrigated and contaminated areas.

The great mobility of the onchocerciasis vector (black flies, particularly *Simulium damnosum*) means that there is a risk of creating areas favourable to its spread wherever dams create artificial waterfalls. Thus the policy of building small dams in Upper Volta has transformed every weir into breeding grounds for *Simulium* spp. and extended the frightful disease of onchocerciasis to valleys where it had previously been unknown.

The construction of the dams needed for irrigation has therefore both positive and negative effects on parasitic diseases. Above the dam, the reduction of the flow and turbulence of the water has an adverse effect on simulid flies but creates favourable conditions for the proliferation of snails and anopheles mosquitoes. Below the dam, the turbulence of water is not favourable to the development of malaria or bilharziasis vectors, but it does provide favourable conditions for the spread of onchocerciasis.

Generally speaking, the "concreting" of an irrigation system constitutes technical progress as well as improvement of health conditions. Ideally, irrigation canals should be closed and, if possible, subterranean. These solutions are costly, but effective, as the example of south Tunisia has demonstrated (Worthington 1977).

Bilharziasis and onchocerciasis are only two examples of vector-transmitted diseases which affect the welfare of human populations in the regions under consideration. There are, of course, many other human ecological factors of importance, and particularly those which relate to nutrition. It is only when all the technical and health precautions have been taken into consideration that irrigation becomes an undoubtable means of development.

Obstacles to the development of rainfed agriculture

The obstacles to the development of rainfed agriculture, where man does not supplement rainfall with any additional water, stem essentially from the short duration and irregularity of the rainy season. More than the total rainfall, these seem to be limiting factors which reduce the range of plants that can be cultivated and at the same time explain the great variation in results obtained within that range. The duration of the rainy season, which is at most four months, decreases rapidly as one penetrates into the more arid zones. Consequently man can use only those crop varieties that have a short vegetative cycle which depends on the relation between the timing of the rains and the temperature distribution. Thus, there is a sharp distinction between the tropical range of millets and sorghums, which require a combination of warmth and humidity, and the more temperate range of wheats and barleys, which are better adapted to the cold season rains of the mediterranean climate.

One of the problems of tropical agriculture is the date of planting, which must be done at the beginning of the wet season when evapotranspiration is high and the risk of protracted interruption of the rains is considerable. Along the borders of the Mediterranean, the most suitable crops are the drought-resistant early varieties of wheat and barley, which require less than 200 mm of water during the period of growth.

However, it is not so much the limited range of available cultivated crops as the irregularity of the results obtained which constitutes the greatest obstacle to development, entailing as it does insecurity of food supplies that is all the more serious in that it affects regions where the population is rapidly growing. In the Near East and Latin America, population growth rates from 1955 to 1970 were 3.6 per cent in low rainfall zones. In Africa these rates were lower (2.9 per cent) but are tending to increase. Every agricultural season is therefore something of a short-term gamble, the results being dependent on the whim of the climate.

In the semi-arid marginal regions there is indeed a very high inter-annual variability both as regards the total rainfall and the number of days of rain. In Senegal, in 1972, the overall deficit was as high as 50 per cent north of a line between Dakar and Bakel, 70 per cent along the coast and 80 per cent in the Cap Vert region. At Kairouan, in the Tunisian steppes, 545 mm of rain were recorded in 1934, and 54.6 mm in 1945. In Chad, at Bol, where the mean over a thirty-year period was 328.7 mm, only 59.4 mm fell in 1972 and 207 mm in 1973.

A host of examples also proves that the inter-annual irregularity varies inversely with the average rainfall. In some cases no rainfall is recorded during a period of several years (in the drought areas of northeastern Brazil, for example). This makes semi-arid regions highly vulnerable, not only in areas bordering immediately on deserts, but also in a wide fringe zone where there is as much as 600 mm of summer rain or 400 mm of winter rain. Beyond this there are still fluctuations but the consequences are less catastrophic.

In addition to the uncertainty with regard to the total rainfall, there is the uncertainty about the date of the first useful rains on which the opening of the agricultural season depends. Thus 300 mm of rain falling heavily in the autumn, followed by a long period of drought, is of no use at all for cereal crops north of the Sahara. The widespread catastrophe of 1945 which affected an area from the region of El Gharb in Morocco to the plains of Tunisia indicates the degree of insecurity in regions which are normally better protected from fluctuations. On the tropical edge of the arid zone, a dry period after the first rains postpones the real beginning of useful rainfall, ruining the first sowings and making it necessary for farmers to take further seed from their reserve stocks.

The growth of crops depends not only on the total volume of precipitation and the number of days of rain but above all on the distribution of rainfall throughout the rainy season. This distribution is at least as important as the total volume of available water, and perhaps even more important. A higher yield very often occurs in years when there is a lower overall rainfall distributed in a way which is better suited to the requirements of the vegetative cycle. As irregularity in rainfall distribution is a permanent factor, man is bound to find agriculture something of a lottery, more particularly in areas near the desert.

This phenomenon also affects valleys in which the permanent availability of water has led man to extend his agricultural activities by making use of floodwater and its effects. Flood-retreat cultivation remains closely dependent on the water supply in drainage basins, which is governed by the rhythm and regularity of rainfall. Except for the major allogenic rivers of Asia (which are well supplied with water from high mountain ranges) and the rivers which rise in the Andes, the floods closely reflect variations in rainfall, which result in a larger or smaller flooded area and determine the size of the cultivable area. The discharge of the Senegal may fall from 4,200 m^3/s to 1,600 m^3/s in its middle valley, flooding 800,000 ha or a quarter of that area respectively. In 1973 the average discharge of the Chari and the Logone for the year was less than a third of normal and the flood-retreat sorghum did not even reach the stage of forming heads. Valleys, as well as interfluves, are therefore liable to unreliable food supply.

In unirrigated farming the water available for plants comes exclusively from the infiltration of rainwater. Soil water balances have so far seldom been measured. The infiltration of water into the soil and its erosive capacity are largely contingent on the irregularity of rainfall and variations in its intensity. Thus, the sudden onset of a storm will result in considerable runoff, as the soils are not sufficiently permeable for all the surface water to penetrate. However, different types of soils produce different effects. The sandy soils of dunes or ergs have a high infiltration capacity (4 to 7 cm/h) whereas vertisols allow a certain degree of retention, particularly in crevices, which subsequently swell, become clogged and temporarily hydromorphic.

The fertility of the soil is less important than its water dynamics (which depends on the structure of the profile) in determining the soil's agricultural potential. Those soils which offer substantial advantages are at the same time most vulnerable to erosion. For example, permeable sandy soils permit deep penetration of water, which limits evaporation. Their moisture content rapidly reaches and passes wilting point, thus promoting plant growth. For this reason, the inhabitants use such sandy soils on a large scale. But the soil-vegetation-cultivation equilibrium is precarious and wind erosion carries away the upper horizons particularly easily in the case of light soils. Furthermore, during the rainy season, the mechanical effect of rainfall results in selective separation of the elements on the basis of grain-size distribution. The loss concerns particularly the colloidal fraction (clay and humus), thus exacerbating the damage caused by erosion.

One characteristic feature of the production techniques used by civilizations inhabiting semi-arid fringes is that crop cultivation and animal husbandry are kept totally separate or are inadequately associated. However, the situation varies widely. In Sahelian Africa, cattle are very seldom used for agricultural work. At the most one may find instances in which contracts have been made with nomadic pastoralists for the supply of manure. In North Africa, the swing plough is drawn by oxen on heavy soils; mules are used in mountain areas, and the dromedary more rarely except in Tunisia south of the main ridge and in Libya. In the central part of the Bolivian Altiplano (a high-altitude desert) pigs are sometimes used to prepare the ground prior to sowing. But there is no real integration of crop growing and stock raising; what occurs is merely a juxtaposition of two activities in the same area, favoured to some extent by the amount of fallow.

The use of a rotation system in which crops and fallow alternate in a two-year cycle, is, indeed, often found in traditional agriculture; this approach is essential for agriculture when there is often no means of fertilising the soil, which is subject to intense erosion. One further advantage of these fallow lands is that they retain the water from two rainy seasons for a single harvest. The few exceptions to this system of crop-fallow rotation occur in very limited areas in the deserts of Chile and Peru. In the valley of the Chilca (65 km south of Lima) old techniques are found whereby extensive excavations called *pukios* are made in the soil (100 x 50 and 30 x 30 cm) in order to retain moisture for the cultivation of maize, vegetables and fruit. The same technique is used in arboriculture (vines and fig trees) on the island of Lanzarote in the Canary Islands.

In semi-arid fringes with winter rainfall, arboriculture is an important activity. In mediterranean zones, tree crops (in particular figs and olives) have a high nutritive value. Figs, olive, almond and pistachio trees do not produce satisfactory yields outside the semi-arid zone. A remarkable dry farming technique has been preserved by the inhabitants of Sfax and has provided a means of developing enormous olive plantations, from the Tunisian steppes to Tripolitania, where the average rainfall is as low as 170 to 250 mm. These results are achieved by a technique of loosening the soil by ploughing and surface cultivation which is unfortunately unknown to the majority of those tree-growers.

As a result of the current population increase and the spread of cash cropping, crops dependent on rainfall are spreading more and more into areas where the yields are largely a matter of chance.

The cessation of the practice of fallowing, the resulting exhaustion of unfertilized soils and their vulnerability to erosion, and the peasants' need to earn money, have induced them to extend their activities to regions in which there is even greater risk. In establishing themselves in the vicinity of wells, they make use of infrastructures previously intended for animal husbandry. Large mechanized farms have grown up in which the tractor has replaced the swing plough for the production of cotton and cereals (northeastern Iran, Iraq, Jordan, eastern Morocco) (Sébillotte 1973). This extension of cultivated areas has taken place at the expense of grazing lands, in spite of the fact that the economic results, which are largely a matter of chance, do not justify this ill-considered occupancy of lands. The increase in cultivated areas has often meant inadequate land use, an obvious consequence of which is the destruction of the fragile plant cover and increasing erosion. This brings on the common process of man-induced desertification, which remains the principal cause of the extension of desert areas.

OBSTACLES TO URBAN, INDUSTRIAL AND TOURIST DEVELOPMENT

The fundamental problems of animal husbandry and agriculture are directly related to the harshness of the environment, but industrial and tourist development are not limited to the arid zone. Whether the climate is arid, temperate or humid tropical, housing, unemployment and traffic problems are mainly urban problems. Whatever the latitude, the study of an industrial market is always subject to the same variables (cost of energy and raw materials, manufacturing cost and distribution). Moreover, urban and industrial growth has had little impact on the standard of living of predominantly rural populations. The real obstacles to development are to be found in economic and social systems and not in environmental constraints. In the desert and its fringes, however, these constraints (heat, scarceness of water and isolation) often appear to aggravate the problems of any city or industrial unit.

Obstacles to urban development

Urbanization in the cities of the arid world has been rapid and spectacular. The very high birth rate and the massive influx from the countryside, augmented by climatic uncertainty and bad periods (such as the 1972 drought in the Sahel and India) have increased migration towards the cities. Nouakchott's population was 17,000 in 1965, 40,000 in 1970 and over 100,000 by 1974. In Niamey, migrants account for 7 per cent of an average growth rate of 9.6 per cent. Bagdad's population went from 800,000 in 1957 to 2,000,000 in 1972; from 3,5 million in 1972, Teheran is expected to reach 5,5 million by 1991. Lima's population increased from 1,8 million in 1961 to 3 million in 1969, and will reach 5 million by 1980. This increase in urban population often results in an uneven national distribution of population. One-fifth of the Iraqi population lives in Bagdad and by 1980, one-third of all Peruvians will live in Lima-Callao.

Urban problems are all the more acute in the desert because, although their demographic expansion is quite recent, these cities are of ancient foundation. The old world produced an early urban civilization based on caravan trade and the expansion of Islam. In Africa, as elsewhere, Islam led to the creation of cities (garrison towns, administrative and commercial centres) clustered around their mosques. The colonization of North Africa and Latin America, along with the exploitation of mining and energy resources, has also changed the old cities and given rise to the construction of new cities.

Today these urban centres face an unprecedented demographic boom. They are poorly equipped, if not totally unable, to house, employ, feed and provide water for a mass of impoverished newcomers. Urban development should not be confused with a proliferation of unintegrated building. The need for water is then far greater.

Thus, an urban way of life, because of the variety of domestic uses, implies an increase in daily water consumption. The minimum consumption per person in a city like Bagdad is 60 litres a day, but in most cities it may go up to 100, if not 200 litres. However, most big cities are good catchment basins: watered foothills (Teheran, Damascus), intramontane basins (Shiraz) allogenic rivers (Kabul, Cairo), coastal areas where desalinization plants have been set up (Kuwait, Saudi Arabia). However, the high cost of these plants remains an obstacle for poor countries.

A specific hindrance to the urbanization of arid zones is the large water consumption in towns (Bagdad consumes 34,000,000 cubic meters a day from the Tigris), and the competition for water between agriculture, industry and cities. Moreover, misuse of water, waste and various losses, have serious consequences and may result in local subsidence where groundwater is not replenished. In the Tucson urban area, sinkage of 254 mm has been recorded.

Depending on economic policies, water may limit urban development. Water needs are also qualitative; having a sufficient amount of water is not enough, since the water must also be drinkable. Water can carry dangerous amounts of mineral salts in solution or solids in suspension. Quite often the water does not satisfy the standards of toxicity set by the WHO. City dwellers refuse to drink brackish water that they rightly consider dangerous. Finally, though some cities have a satisfactory water system, only a few possess an underground sewage system and only exceptionally do they have a sewage treatment plant. Poor sewage systems are a great hazard to public health.

The effect of heat on people is a hindrance to urban development. Hyperthermia, insomnia, fatigue and problems of adaptation may result from poorly designed houses. The effect of huge solar radiation on buildings used to be overcome by simple but relatively efficient techniques in traditional dwellings (uses of shade, natural ventilation). As a consequence of the demographic boom, and anarchic urban growth, these dwellings have been replaced either by totally unadapted so-called modern constructions or shanty towns built with second-use materials unable to protect man from heat. Other climatic problems should be taken into consideration: atmospheric inversions characteristic of the lower troposphere in arid zones encourage the concentration of pollution above cities (industrial pollutants and automobile exhausts).

Not only are these cities lacking in communal facilities and adequate housing, but in certain cases the structures are unable to meet the requirements of modern urban life (e.g. industrial and car pollution).

The legacy of history has made the urban area extremely difficult to develop. The central sections of Muslim towns, criss-crossed by a labyrinth of small streets, are generally antiquated and poorly ventilated. These difficulties are often aggravated by topography (hills and strategically positioned rock sputs). These highly-populated districts are the most dilapidated and the most in need of renovation that should take into account the positive aspects of the traditional architecture. Towns established by colonial governments have a dual structure. Around a nucleus of modern buildings often with a geometrical ground-plan, clusters of chaotic and overpopulated peripheral districts, totally lacking in public utilities, have been organized or have sprung up spontaneously.

Finally, the urban thrust has spread cities out disproportionately: Bagdad is 30 km long and Teheran, now over 25 km, should cover 100 km by 1991. It is becoming more and more difficult to organize transport, and equipment costs have skyrocketed. The rise of land prices in the centre increases segregation by removing the underprivileged majority to the periphery.

Obstacles to industrial development

Although the uneven distribution of natural resources creates an imbalance between nations which have fossil fuels or basic ores and those which do not, some obstacles to industrial development are common to all countries of the arid and semi-arid world.

Climatic conditions, scattered plant cover and surface outcrops of rocks, facilitate discovery and exploitation of minerals, usually without causing serious damage to the environment. Excessive heat, however, may have considerable effects on man's working capacity.

The first effect is dehydration. Human water requirements are estimated at 10 litres a day for strenuous work at a temperature of 30 °C. A great tendency to sweat is accompanied by a quickening of the heartbeat and a great loss of salt. Dilated blood vessels may cause circulatory disorders, and a lessening of attention and slower reflexes also occur.

It has been shown that in the Sahara the curve of accidents in the oil industry closely follows the curve of temperatures. Studies of the deterioration in tasks requiring vigilance have shown that above 34 °C the number of mistakes increases exponentially with the temperature. These climatic constraints, coupled with endemic disease and illnesses connected with the extension of irrigation, constitute a very real obstacle to development.

Water resources, and particularly the availability of large quantities of low-cost energy are two of the basic conditions of industrialization. Modern production techniques use large amounts of water: 600 m^3 for a ton of nitrate fertilizers, 1300 m^3 for a ton of aluminum, 150 to 300 m^3 for a ton of steel, 3 m^3 to refine 150 litres of oil. However, because of the extreme scarcity and variability of rainfall, water tables in the desert are inadequately recharged. In the Kalahari, for example, there is presently practically no present formation of deep reserves, and water tables are recharged on average once every 10 years. In semi-arid regions infiltration is more frequent, but fluctuations in water table levels are an important obstacle to the establishment of industrial activities which need a regular water supply. The development of steelworks at Monterey (Mexico) for instance, has been hindered by progressive depletion of water tables as a result of deep pumping. To remedy this it will be necessary either to divert the waters of the Rio Grande, or even to create a desalination plant and an aqueduct.

Furthermore, the salt content of runoff water is generally high and varies according to the nature of the rocks of the catchment basin: thus, magnesium chloride is found in the Dead Sea, gypsum and sodium chloride in southern Tunisia, sulphates in western Pakistan and nitrates in the Chilean desert.

When the geological structure is favourable, but a deep water table is lacking (northern Sahara, southern Africa), or where water cannot be brought in from the outside (coastal deserts along the edge of the Andes), scarcity and the quality of available water makes necessary, in most cases, industrial desalination (by ion exchange, electrodialysis, inverse osmosis, or distillation). However, the cost of water obtained in this manner, including fresh water obtained by desalination of sea water, remains prohibitive. What is more, desalination, which may be viewed as a long-term solution to the problem of water, requires large quantities of energy, the high cost and scarcity of which are an important obstacle to industrial development. In Afghanistan, the low level of energy production and the lack of a national distribution grid are the main factors limiting industrialization. Oil producing countries have also been faced with such problems. In Saudi Arabia, electricity remains rare and expensive and its transmission very difficult.

Industrialization cannot be conceived of without an effective communications network. The railroad has long been a sure way of overcoming distance as well as aridity (the diesel engine does without water almost completely). Though topography is generally favourable, with the exception of the high altitude Andean deserts, soil conditions are very varied (extended limestone hardpans and regs are very stable desert surfaces, but dunes and sandy stretches are harder to cross). During the construction of the Transcaspien, to cross the Kara Koum, it was necessary to establish cuttings, consolidate embankments with wattle walls and plant *Arthrophytum persicum* or *A. haloxylon*; these solutions are, however, not applicable everywhere, particularly in the Sahara.

But the main obstacles are economic. A communications network obviously will not be profitable unless it has sufficient goods to transport. With this in mind, such a network is conceivable only for mining. The railway which runs from F'derik to Nouadhibou is profitable only because of the long, heavy, and consequently very slow freight trains which use it. Unequal geographic distribution of resources in the countries of the arid and semi-arid world adds to the complexity of the situation. The Arab countries of the Middle East, for instance, possess over 60 per cent of identified oil resources; in the same manner the income of Chile and Mauritania is derived almost exclusively from copper and iron. In contrast, Mali, Chad and Sudan cannot at present hope to export energy or industrial raw materials.

The problems of industrial development cannot be considered in the same terms for these two groups of countries. In addition, oil producing countries are extremely varied. Those with a high per capita income (Abu Dhabi, Kuwait, Libya, Qatar, Saudi Arabia) have rapidly increasing financial resources. Their populations are however small. As a consequence they suffer from a shortage of industrial manpower and a limited domestic market. Their transport infrastructure is underdeveloped and basic ores are scarce, or as yet unexploited. On the other hand, in those oil producing countries with a lower per capita income (Algeria, Irak, Iran), conditions are more favourable to industrialization. The labour force is larger, the communications network is well-developed (but in some cases still insufficient, as in Iran) and the mining resources are more diversified and often already being exploited. Whatever their situation, most of these countries already having mining industries and produce a variety of consumer goods destined to avoid, and progressively replace imports (textiles, light engineering, building materials, for example).

These obstacles are made still worse for countries with no outlet to the sea. Remoteness of ports, the length and difficulties of communications, seriously add to transportation costs. The borders of Niger, for example, are 650 km from the ocean, and the distance from the Chad to the Gulf of Guinea is over 1,500 km and is covered successively by road, rail and sometimes waterways, as between Pointe Noire and N'Djamena. The effects of isolation and great distances are felt all the more because population is unevenly distributed and it is often necessary to cross empty or underpopulated areas between high density regions, as in Mali. The landlocked situation of these countries, which as a consequence unfavourably affects the balance of payment and discourages investment, remains one of the major obstacles to development.

Tourism

The physical and climatic characteristics of arid and semi-arid zones are particularly favourable for the development of tourism, given a climate suited to winter tourism, unusual landscapes, in some cases outstanding archaeological sites, attractive coastal regions and the possibility of establishing national parks with little difficulty. Based on these factors, several countries view tourism as a basic industry and even as the vehicle for their overall development. The development of tourism has several positive effects that may theoretically benefit other sectors of the economy: improvement and faster economic return on roads and telecommunications, increase in the number of hotels, large investments in public health and the provision of drinking water, creation of jobs. Thus, in order to satisfy the growing demands of a leisure civilization in the industrialized countries, the development of tourism in the arid and semi-arid regions is likely to be very rapid.

This development encounters, however, a number of obstacles linked to economic fluctuations, international fashions and outside financing. Tourism is by nature highly dependent on foreign demand. It is thus subject to international competition and the fluctuations of a seasonal market. It is also remarkably sensitive to political uncertainty. Economic studies in Tunisia have shown that tourism entails the import of expensive finished products, estimated at between 14 and 27 per cent of gross profits. This tends to reduce the inflow of foreign currency. In most cases the State takes part in the financing of these projects (directly or indirectly by granting payment facilities or subsidies) even if doing so means running into debt. Moreover, the seasonal nature of tourism ties up capital in luxury facilities in generally poor countries. The development of tourism along huge coastal areas may sometimes deprive farmers of valuable lands. More serious are the effects on water supply; in an environment where water is not readily available, tourism is a new competitor. Moreover, the multiplier effect of tourism, very considerable in some industrialized arid countries, requires a specific socio-economic context. Finally, alongside an industry geared to the needs of foreigners, it is desirable to develop recreational resources primarily for local populations.

III. DEVELOPMENT PROPOSALS

Two extremes should be avoided in all development planning:
- *passé* romanticism which calls for a return to traditional ways of using arid and semiarid zones and which would quickly result in the transformation of these areas into natural parks or show places of traditional civilizations for more or less informed tourists;
- a scientistic and futuristic point of view which, under the guise of preparing for the 21st century, pretends that all problems can be solved by science and technology and concentrates on fundamental research rather than on the introduction of practical techniques likely to help satisfy the immediate needs of the local population or to improve their situation.

While the possibilities offered by future technical discoveries should not be ignored, care should be taken to avoid too rapid a break with tradition. Any change in the symbiotic relationships and balances between human societies and the natural environment calls for careful reconsideration of the whole. A distinction must also be made between what is hoped for and what is in fact possible, as well as between science fiction and science which can be applied in the real world. Finally, some institutional frameworks have to be reviewed insofar as they are obstacles to development (these institutional obstacles have been mentioned under agriculture and animal husbandry). Education and training are obviously important to the development of these zones, but these are essential factors in any development project and their importance has already been stressed elsewhere.

PROPOSALS FOR ARID ZONES

In arid regions, human settlements can be established only under special hydrological and hydrogeological conditions (excluding extractive mining, where water is brought in at great cost). Large urban areas such as Lima were able to develop because of the feasibility of bringing in water and because of their location not far from the sea. The problems of these large urban areas are fundamentally the same as those of urban areas in other climatic zones, in particular as regards the desirability of developing smaller-sized urban units (through a policy of medium-sized towns).

The crisis of traditional oases in the Old World is more unusual. With few exceptions, oasis agriculture can no longer compete with modern sector jobs. A modest civil service or military pension, not to mention the wages paid for public works projects or for work in oil fields, provides a higher income than that obtained from agriculture. Many oases of the northern Sahara, such as that of Ghadamès in Libya, owe their survival to money sent in from outside. In some circumstances, the date-palm which produces high quality fruit of the "deglet nour" variety, could be the key to modernisation, but it requires much care and manpower and is subject to attacks of "bayoud", a fungal disease about which little is known.

In modern or modernized oases, special pricing procedures should be used for water intended for agricultural purposes, since there is a risk that other users who are able to pay much higher prices will monopolize all the supplies for their own benefit. Thus, in Tunisia, water is used by tourists to the detriment of arboriculture; in Lima, the amount of water used domestically by city dwellers leaves little for gardens. Techniques for recycling water and for using it in agriculture become necessary.

Animal husbandry in arid zones is associated with a way of life that seems largely doomed. Its survival is linked to that of certain institutions whose revival through economic development is unrealistic since it would deny pastoralists the only remunerative function they formerly enjoyed, that of caravaneer or "desert guides".

PROPOSALS FOR SEMI-ARID ZONES

Development of animal husbandry

Modernizing pastoralism, improving its productivity, ensuring its integration into the national economy and its harmonious existence side by side with other activities - i.e. turning it into a developing sector while it competes for land with crops without, however, escaping climatic risks - all this demands difficult technical decisions and tenacious political will. Furthermore, plans are not all equally sound. Although important results have been achieved in prospecting for and extracting water, particularly groundwater, and an increase in livestock has been brought about through advances in veterinary medicine, little progress has been made in improving rangelands.

The first requirement in developing animal husbandry is to ensure that herds are provided with more ample, regular and nutritive food. As they feed on the natural vegetation of the range, the main objective should be to ensure its rational use. To do so involves, first of all, limiting animal numbers. However, the causes of overgrazing demonstrate the difficulties of an operation which can be successful only within the context of an overall policy for land use management and integrated economic development. This includes the search for ecological solutions such as that within the framework of Project 3 of the Man and the Biosphere Programme (MAB) of Unesco and of the FAO programme on Ecological Management of Arid and Semi-Arid Rangelands in Africa and the Near and Middle East (EMASAR); the preparation of land use maps such as the Soil Map of the World (Unesco-FAO, in publication since 1971).

Rational use of rangeland also implies a change in management. It is precisely by means of a land management policy that this can be achieved. The use of pasture has too often been disrupted by the fragmentation of pastoral societies, the weakening or even disappearance of their old organizational structures, and the encroachment of farming. Any attempt to restore these pasture lands must therefore involve the allocation of land specifically for the purposes of animal husbandry, where in particular the movement of flocks and herds could be rationally organized. It also involves the allocation of this land to coherent pastoral units, capable of management, which might take various legal forms from the simple association of traditional groups to a co-operative or a company. Experiments in Kenya and Syria show that land division combined with the allocation of sectors reserved for livestock to the responsible producers' organizations is not an abstract idea but a practical one. A measure of this kind would also prevent investments (particularly water points), intended in principle to benefit pastoralists, from being used for other purposes, so that the pastoralists rarely use them.

It is also within the framework of precisely defined pastoral areas that measures of a third type can be introduced to improve rangelands. This would involve first of all limiting livestock numbers, particularly in the steppes and savannas where the main problems of overgrazing occur and where the over-concentration of cattle causes the disappearance of the best forage species, as has been shown in Somalia. In addition, grazing lands would be used in rotation, as in certain societies of Africa and the Near East, by setting aside areas grazed only at the end of the dry season and spontaneously regenerated by eight to ten months of rest during the rainy season. As regards direct improvement of natural rangelands, few feasible proposals can be put forward because of their diversity (linked to the nature of the soil, species composition, plant density, nutritive value, biology, etc.) but also because no research has been done in this field.

The importance of trees as animal feed has very generally been underestimated. A number of protective measures could be instituted to promote rational use and many species could be increased by systematic planting (particularly leguminous species) to provide nutritive fodder reserves for the dry season. A systematic policy to encourage fodder trees could be a keystone of land management programmes. In the case of animal husbandry, this would not only solve the problem of fodder reserves for the end of the dry season, but would also provide a use for the land around water points, which is abandoned most of the year for lack of pasture.

As regards the improvement of annual pastures and, in particular, the enrichment of fallow in cereal crop rotations, it is hardly possible to put forward any concrete proposals. Although various authors have referred to possible fodder and leguminous crops, it would seem that the only example is outside the Third World in southern Australia where fallow ground is being transformed into improved rangeland by the introduction of subterranean clover. This is a virtually unexplored area of research, at least in those regions where rainfall falls below 500 mm, since it seems that *Stylosanthes* and *Medicago* give satisfactory results only when annual rainfall exceeds 500 mm.

Consideration should be given to the rational use of fodder crops on irrigated land. So far their rôle has been extremely limited because of the high population in these areas and the low price of meat, which explains why the few examples of irrigated lucerne fields are found only in connection with dairy farming as in Syria or Chile.

In the matter of water management, a wealth of experience and a wide range of proven techniques are available. It is on the rational management of watering places that stress should be laid. The experience of the African Sahel highlights the danger of high-discharge boreholes. By encouraging abnormal concentrations of livestock they risk leading to overgrazing which rapidly destroys the plant cover and renders the water points themselves unusable. Whenever possible, a policy based on clusters of rationally sited wells is preferable to the systematic sinking of deep boreholes. Above all, it is vital that any water management policy is accompanied by a land management plan so that the numbers of livestock using the wells or boreholes are kept in line with the number that can use the rangeland that they supply. This implies drawing up a policy for water and rangelands at the national level in collaboration with the pastoral societies; the demarcation of grazing areas could provide a framework for such a policy. These areas should be entrusted to groups responsible for their management; ownership of such areas needs to be clearly defined at the outset.

Rational management of the land is, moreover, an essential requirement of modern methods of livestock production, particularly as regards regional specialization. Considering the diversity of the countries in question and the very unequal place of arid and semi-arid regions in them, it is futile to attempt to draw up a general model for such specialization. An analysis of the most encouraging spontaneous initiatives taken to remedy the current low productivity of herds, suggests that dry countries with nomadic animal husbandry should concentrate on breeding and establish relations with livestock fattening countries and areas in either moister regions, where farmers are often able to fatten livestock in cattle sheds or on their land, or in areas where irrigated land is used for this commercial enterprise.

This latter solution is at present hard to apply in non-industrial countries for economic reasons, especially the unsatisfactory prices obtained for meat and even more so for live cattle. Thus the question of outlets and market organization is a vital one. It involves national policy concerning the relations between the towns and developed regions on the one hand and marginal country areas on the other, and relations between countries affected by arid and continental climates and those countries where more favourable natural conditions prevail. One important difficulty is, however, the exploitation of the breeding areas by livestock fattening specialists. Examples show that live cattle are purchased for next to nothing at the gates of modern industrial fattening installations and that it is the businessmen who make the most profit on the operation. The most satisfactory technical solutions are ineffective for development unless they are integrated into an organizational system which ensures that the producers benefit.

Development of irrigated crops

Rational management of water at the national level is of particular importance in arid countries in order to prevent waste and to better regulate the use of water in various sectors. A central agency responsible for such water policy could achieve this objective, as shown, for example, in Hungary and Israel.

In any case, the development of irrigated areas should always be based upon thorough studies of the quantity and quality of the water available. When a watercourse flows from one irrigated area to another or from one State to another, there should be agreements which define the volume of water as well as certain quality standards (salt content, amount of suspended solids, etc.). Furthermore, groundwater should be carefully analyzed before it is used for irrigation. Studies must be made of changes in time and space of those fossil groundwaters which are little replenished if at all. Analog models, such as those made of groundwater in the northern Sahara and in the Chad Basin, may be used for the exploitation of other water systems. The social and economic effects of using this water must also be foreseen.

Irrigation projects should be integrated within a framework of complementary and interdependent activities, in the context of a particular drainage basin. Development should plan overall land management and should take account of the way the land has been traditionally used.

The building of a dam is not in itself a sign of development. For a dam to be effective, its impact on the whole valley, upstream as well as downstream, must be taken into account. If the necessary measures to protect the soil from erosion are not taken upstream, the storage lake is liable to become silted up more rapidly. If provision is not made downstream for drainage, vast areas may quickly become sterile. The regulation of the amounts of water to be made available for irrigation purposes must be calculated according to the climatic data, the results desired from irrigation and the quality of the water being used. Under certain soil conditions irrigation with water relatively high in salt content (3 to 4 g/l) is possible if accompanied by adequate drainage. Studies carried out on this subject, particularly in Tunisia, could be useful to other countries.

Enormous dams can create more problems than advantages. Thus for each individual case, all possible alternatives should be carefully examined, such as, for example, a series of smaller dams. Until now effective techniques for reducing evaporation in reservoirs have not been found and as a result there is a considerable loss of water in storage lakes (nearly one out of eight litres in Aswan, for example). On the other hand, the great volume of water which these dams can hold make it possible to mitigate the effects of prolonged droughts.

The economic and social management of irrigated areas is just as important as their technical management. If allocations of irrigated land are to be made, care should be taken to make plots which are large enough and capable of producing surpluses which will ensure the repayment of loans. The financing plan must allow for associated installations to proceed at the same rate as the main engineering works. Delays in the former operations often lead to disastrous consequences both for agriculture and health, and ultimately from the financial point of view as well, because the investments already made do not yield satisfactory returns. Adequate operating funds should be provided from the outset for the rapid establishment of the secondary installations and for their permanent maintenance.

An appropriate choice of the dimensions of the irrigated areas is essential to ensure a suitable return on investment. A recent investigation (Worthington 1977) based on reports from the national committees of the International Commission for Irrigation and Drainage of 29 countries and involving 91 irrigated areas (inside and outside the arid and semi-arid regions) shows that the excessive complexity of many water management installations often causes water losses far superior to those resulting from infiltration and evaporation in the irrigation canals. The best results are on two different scales. The optimum size for rotation units seems to lie between 70 and 300 ha. Above 600 ha and below 40 ha, the management of available water generally suffers marked deterioration. The optimum size for irrigated areas seems to be between 3,000 and 5,000 ha and the management of available water deteriorates markedly above 10,000 ha and below 1,000 ha. As a consequence, it has been suggested that units be created no smaller than 1,000 ha and that large irrigated areas (over 10,000 ha) be broken up into blocks of 2,000 to 6,000 ha.

These conclusions are close to the previously expressed idea that the gigantic proportions of certain projects do not necessarily guarantee success and certainly are not the most effective way of using the available water. Within each block the topographic conditions, as well as the size of the different farms, obviously determine the limits of each rotation unit. However, it seems important that within different units, the rotation of water supplies to farms or groups of farms be independent from the distribution of water in neighbouring units, in order to avoid losses of irrigation water as much as possible.

Finally, the success of irrigated areas, in the last analysis, depends on the type of farm that has been established or encouraged. At this level solutions may be very varied, according to historical and demographic conditions or political options. In Australia the extension of irrigation in the Murray basin has been hampered by a shortage of manpower. In other countries, on the contrary, it is difficult to make room for all (in North Africa for example). Different types of farms (large capitalist farms, farms managed by a state administration, small independent traditional farms, farming of "social" lots more or less well connected to co-operatives, etc.) may be found side by side within the same irrigated area, as in Morocco.

A recent study of the modernization of agriculture in the Dez area of Khuzistan in Iran (*op. cit.*) has emphasied the difficulty of obtaining results that are economically, as well as socially, satisfactory since mechanization increases unemployment. The establishment of agricultural enterprises, managing between 10,000 and 20,000 ha has brought about a substantial increase in productivity, but has reduced labour requirements by 75 per cent. Agro-industrial complexes, like the one in Haft Tappeh, are more advantageous in this respect. Although the return on investment is much lower, they employ a large labour force. This labour force, made available by the mechanization of agriculture, has been employed locally in sugar cane processing plants. Food industries, and more generally the industries involved in the basic processing of agricultural products, are able to offer a certain number of jobs to farmers left jobless as a result of the widespread use of machines and the regrouping of land. Thus the establishment of agro-industrial complexes seems highly desirable, both for economic and especially for social reasons.

The same study reveals that the production of marketable surpluses by traditional farmers in the irrigated areas does not call for the disintegration of traditional social communities, or the suppression of all pre-existing agrarian structures. Although basic foods such as cereals are sometimes more profitably cultivated on large plots with powerful machines to reduce production costs, other crops may be reserved for traditional small farmers who are perfectly capable of developing their production, providing they receive fair remuneration and adequate technical and commercial support. This may be achieved by a network of co-operatives, the effectiveness of which will depend on an acceptable compromise between two extremes: very large co-operatives, powerful because of their organization and their impact on the economy, but showing little concern for farmers' aspirations, and small co-operatives, closer to the farmers and controlled by them, but lacking the necessary size and sufficient means to influence other economic and social organizations.

Another interesting experiment (Balbo 1975) now in progress in some Sahelian countries (Mali, Senegal, Chad), seems to indicate, at least in some cases, that large irrigation projects and those carried out on a small scale in traditional villages, are not incompatible, but may supplement each other. Large projects (harnessing large allogenic rivers) contribute to long-term development, particularly in the Sahel. However, these large projects will be successful only if many precautionary measures are taken in due time. Moreover, their construction costs, as well as their operating costs, are very high and the highly advanced management techniques, which large projects necessitate, limit the participation of the local population.

In small village-level projects (irrigated areas between 10 and 25 ha supplied by water pumps, for example) water management is far from optimal by modern standards. These projects offer, however, a great many advantages. As construction and maintenance costs are low, construction itself can be very rapid. This may be important in view of the urgent problems of some densely populated areas. The demonstration value of these projects is also very important, since in the early stages, costly imported techniques are not indispensable, and local labour may be employed in all operations without disrupting farming communities. Large scale local participation seems to guarantee success, since people more willingly and more effectively maintain works which they helped construct. Finally, new habits and new skills may be acquired in this process, and these basic units may eventually be incorporated into larger projects with greater means at their disposal. Thus it appears that large modern projects and small village-scale operations can complement each other, provided the environment is suitable. While awaiting their incorporation into larger programmes, small projects teach irrigation techniques and involve farmers much more in their own development.

However, it would be foolish to suppose that irrigated areas are able to support large populations while providing an adequate standard of living for all their inhabitants. It seems particularly important to avoid creating pockets of prosperity that attract large migratory movements from underprivileged regions. Efforts to plan and modernize the economy should be extended to regions in their entirety and not confined to irrigated areas alone. Finally, at a higher level, the marketing of agricultural products calls for an effective communications network to the vital centres of a country or of other countries.

In a few cases where large-scale investments are not limited by a need for immediate financial profit, agriculture by irrigation, based on the most advanced techniques and yielding excellent results, can be instituted in a small number of areas. Such undertakings, infrequent but not impossible, should nevertheless fall within the framework of an integrated land use policy.

Farming in irrigated areas requires particularly demanding and specialized techniques (very careful working of the land, crop protection, etc.) which cannot be improvised. Its introduction into regions where it is unknown requires a difficult transformation of traditional attitudes (the transformation of nomadic herdsmen into farmers, or of simple farmers into irrigation farmers has often met with failure). An example among others is provided by the development of the area of Khashim el Girba in Sudan. This project, started in 1964, was designed to resettle Nubians whose lands had been covered by the waters of Lake Nasser, upstream from the Asswan High Dam. Another objective of the project was to settle on irrigated land nomadic and semi-nomadic populations, which up to that time had moved around the region. The unsatisfactory results obtained so far are attributable in part to technical difficulties, but mostly to adaptation problems. The various populations transferred to the area had not been sufficiently prepared for their resettlement. This gave rise to high absenteeism (the Nubians still had permanent interests in some towns and the nomadic and semi-nomadic populations were still attracted by livestock rearing) not compatible with an effective development of irrigated agriculture.

Thus, irrigation requires not only technically trained personnel with a sound knowledge of agronomy, but also people who will do the job effectively. Errors are frequent and mistakes may have serious consequences. The need for constant supervision of changes in land and water calls for programmes in education and information to accompany technical operations.

Popular health education is also an essential part of all measures of prevention of parasitic diseases associated with the expansion of irrigation, as well as of eradication programmes of such diseases, such as the onchocerciasis control programmes undertaken by WHO.

It must be remembered therefore that irrigation, both in itself and in the associated measures it calls for, involves techniques which require very careful handling. Any undertaking in the field of irrigated agriculture must be preceded by painstaking, interdisciplinary surveys. Pre-project studies must start at the level of the individual farm and the individual farmer.

Account must evidently be taken of climatic possibilities in choosing the crops to be grown. A right balance has to be struck between food crops and cash crops; the marketing of the latter must be organized in a satisfactory way, avoiding in particular the system of selling standing crops. It is usually desirable to reserve at least part of the irrigated land for growing fodder crops. This permits better integration of agriculture and animal husbandry, since this integration remains an essential component of any agricultural "revolution".

Close association between crop production and livestock raising is mutually beneficial. Better fed animals are more productive; the fertility of the soil, rapidly exhausted by irrigation, can be regularly improved by the addition of manure. Seasonal irrigation and year-round irrigation pose their own special problems. The transfer from the first type of irrigation to the second, which often appears as the ideal solution when water resources are sufficient, is not always desirable (due to the danger of waterlogging and salinization of soil, and of soil exhaustion). On the other hand, the advantage of alternating dry farming and irrigation farming on the same plot when the climate permits must be stressed. This method allows for less and better use of water inputs; in addition, short-term drying out helps restructure the soil.

Development of rainfed agriculture

The agricultural production of countries with low rainfall is of vital importance to their economy. A report by FAO estimates that 37 per cent of the GNP in these areas comes from agriculture which employs 67 per cent of the active population. In addition, the areas under cultivation are expanding. In North Africa and the Near East, the area under cereals increased from 10.6 million ha in the 1950's to 11.7 million ha in 1970 and in the Sahel-Sudan region, from 16.3 million ha in 1960 to 17.2 million ha in 1970. Very considerable increases have been noted in the Sudan and Ethiopia.

This expansion has been achieved at the expense of rangelands and has only accentuated the decline of pastoral resources, while the results (low cereal yields of 700-800 kg/ha) do not compensate for the harmful effects. Production lags far behind the needs created by the considerable growth of the population. This is why technical progress and increased production, which are vital to development, require the stabilization and limitation of areas under cultivation. The answer is not so much to cultivate larger areas as to improve yields and alter techniques. There are two possible courses of action: either to establish large farms capable of making profitable use of modern means of production, or to spread knowledge of new techniques by initiating activities among all small farmers. Since one solution does not rule out the other, a few ideas are put forward here which could be applied in the immediate future by farmers in semi-arid zones. Vast world-wide programmes have been ruled out, since experience has proved them utopian.

Development of dry farming techniques

The first improvement could be made in the varieties grown, the rotation of crops and the elimination or the reorganization of fallows according to specific circumstances.

By using native, hardy varieties already selected by centuries of peasant experience, it is perfectly feasible to obtain plants which are resistant to drought and which adapt to the soil's water content, giving a minimum yield in unfavourable periods and excellent crops when rainfall is abundant. Varieties of sorghum are already known in tropical areas (the *Fédérita* variety in Sudan). High-yield varieties of wheat have produced good results in Mexico. Considerable work should be done on seed-bearing legumes (lentils, chick-peas, beans) which enrich the soil with nitrogen and whose nutritive qualities (richness in proteins) are well known. These improvements should be extended to cash crops, oil seeds and fibres (short-cycle groundnut, cotton without gossypol).

The spread of these varieties, combined with the introduction of seed-bearing legumes, can eliminate the need for fallows in regions receiving winter rain. In this way, by appropriate rotation, cereal yields can be substantially increased, and the crop residues produced could be used to feed animals or be sold. Australian experience with legumes of Mediterranean origin (e.g. *Trifolium subterraneum*) could be of great use in regions of the Near East and North Africa having sufficient water supplies. In those regions which receive summer rain, the question of fallow land is much disputed. The development of cassava in some parts of the African Sahel is nevertheless striking (Senegal, Niger). The food value of this plant, which is, paradoxically, of equatorial origin, and the considerable contribution it could make to animal husbandry, deserve attention and publicity.

Yields can also be improved by the spread of simple and inexpensive measures such as the disinfection of seed and the storing of grain in granaries protected against rodents. It is reasonable to suppose that some techniques of cultivation which take advantage of available humidity could immediately be made widely known, for example preparation of the ground before the rains in order to ensure good water retention; spacing and depth of sowing to facilitate the start of growth; and weeding after the plants have sprouted. Use of fertilizers and herbicides calls for a sound knowledge of the quantities to be used. The rise in the prices of these products, however, should lead to great care in view of the heavy indebtedness of many farmers.

Agricultural planning in dry regions, especially the choice and distribution of rainfed crops, should be based on a more systematic use of weather information. In this respect, the results of work already completed in agroclimatology should be better disseminated: transfer and publicizing of knowledge in this field should be more effective. In some cases agroclimatological studies could also contribute to more rational choices between rainfed and irrigated crops.

Systematic use of trees

To combat erosion and stabilize the areas under cultivation, the spread of simple anti-erosion techniques is perfectly feasible. An experiment conducted between 1966 and 1971 in Ader-Doutchi (Niger) shows that dividing plots of land by small stone walls following the contour lines or rows of *Andropogon gayanus*, in conjunction with shallow contour ploughing, has led to considerably higher yields (97 to 192 per cent for cotton, 60 to 80 per cent for groundnuts). In addition, the soil was well conserved during and after the rains. This example shows that modest but effective results can be quickly achieved for the farming community without recourse to large-scale "luxury" projects. The various techniques for soil conservation and restoration, including soil fixation by plants from more arid environments (for example, Saharan plants introduced into degraded arid regions), should become widespread. More costly or more elaborate procedures (such as mulching through use of hydrocarbons and bituminous residues) are of course possible in some cases.

Another way to combat erosion is the systematic use of tree cover, in which certain species are of the very greatest value in soil stabilization. The best example is provided by the *Acacia albida*, whose inverted vegetative cycle affords three remarkable advantages. As it sheds its leaves in the rainy season, the *Acacia* provides the soil with considerable quantities of organic matter. Organic matter in fact doubles under *Acacia* and is still perceptible at a depth of 120 cm. There is very considerable phosphorus enrichment, as well as increase in nitrogen, exchangeable bases and fertilizing elements. Improvement of this kind from trees, is equivalent to annual application of P_2O_5 fertilizer, to normal CaO liming, or, where organic matter and nitrogen are concerned, to the spreading of a hundred tons of fresh manure. A second advantage is that the tree does not cast any shade during the crop season and promotes the growth of the plants beneath it. In the dry season, its foliage protects the soil from both wind and sun. Lastly, its pods and leaves provide excellent fodder for livestock at the time of year when pasture becomes insufficient. By doing away with long bush fallows, the use of *Acacia albida* also provides opportunities for associating animal husbandry with crops, while at the same time stabilizing the terrain on improved, fertile ground. By providing more protection for the plant cover (protected areas, organized rotations, and strictly supervised stock movements) many more instances of the usefulness of vegetation could be observed. In this way, trees could be chosen, in consultation with the farmers, to be used in the new systems of farming. In addition to the Sahelian *Acacia albida*, other interesting species of trees and shrubs are *Acacia aneura* of Australia, *Prosopis tamarugo* of Chile, *Prosopis cineraria* of the Thar desert, several species of *Atriplex* in North Africa and Chile, etc.

More generally, it is necessary to increase the protection of "wooded" areas and to encourage their spread because of the important role played by tree and shrub cover in the protection of the environment and the feeding of livestock. In some cases, windbreaks and forest green belts help improve ecological conditions.

Since environmental conditions are precarious and, in particular, precipitation is seasonal and erratic in distribution, it is difficult to distinguish between "forest", "bush" or "rangeland" (as in temperate regions, for instance) because trees, shrubs and grasses are closely inter-mixed and ecologically interdependent. This close interdependence of various vegetation forms is epitomized by the fact that in drier conditions the vegetative period of herbaceous plants becomes steadily shorter as aridity increases and such plants occur mainly under the shelter of trees and bushes.

The second cause of difficulty in segregating lands for single use development is an economic one. Due to unfavourable environmental conditions, the natural vegetation is open, ranging from woods to scattered, isolated trees and bushes. This renders development and management of vegetation for timber production uneconomic on its own. At the same time, due to the erratic precipitation, average crop yields are not only low but also subject to wide fluctuations. During periods of prolonged drought it is only the woody vegetation which survives and provides forage for both domestic livestock and wildlife. Thus, forest management needs to recognize the dependence of man (and of his livestock) for food, wood and fibres and for other "social values" on the same area of land, whether it is called forest, or rangeland.

Land development strategies, including forestry development, need to aim at "total production" through integrated development and multiple use of all available resources. Such an approach, however, demands a new land management methodology, based on inputs from various branches of the physical, biological, social, economic and managerial sciences, which is much more than the mere co-ordination of inputs by the traditional major land-use disciplines of agriculture, animal husbandry and forestry.

Similarly, fruit-tree cultivation can be greatly extended in regions of winter rainfall. The techniques of dry farming used near Sfax have led to a diversification of production which avoids the risks of monoculture. Thus, almond trees can be planted together with pistachio, fig and apricot and sometimes peach trees. Recent successes of fruit growing in Afghanistan prove that such possibilities exist.

Organization of the countryside

The spread of new techniques means little unless it forms part of an overall development policy. Governments must be aware that agricultural progress depends upon land tenure policy, which must discourage itinerant forms of cultivation. There must be a clear definition of how land is to be used and of the rights and uses by different populations. This implies recognition of traditional grazing rights and the prevention of haphazard extension of cultivation. While controlling the use of land, thanks to flexible structures adapted to specific social conditions, the State could thus encourage technical progress by working with the existing rural population or by launching large-scale pilot projects; the two are not incompatible. These principles have already been recognized by certain countries (collective grazing lands in Syria, group ranches in Kenya, law concerning State lands in Senegal).

This land tenure policy must, in some cases, be accompanied by a thorough change in legal structures. How can share-croppers give enthusiastic support to a development policy when the bulk of what they produce is monopolized by absentee landlords? Any development effort requires a collective awareness which presupposes that the farmers must be working first of all for themselves. Land tenure policy is only one item in an agricultural policy which should provide effective organization for producers by such means as fixing a minimum price for the main products, the organization of credit facilities, and the spread of light mechanization. In many cases, failures can be traced back to the lack of any such organization.

While it is technically possible to increase agricultural production in semi-arid zones, it would be illusory to suppose that increasingly high population densities can be maintained. Governments should think in terms of easing the pressure on cultivated land in order to give rainfed agriculture the best chance of success. A policy of migration or of settling people outside the cultivated areas can be based on the establishment of agricultural industries (meat industries, flour mills, skins and leather works) or on the manpower requirements of centres elsewhere. Such migrations should form part of a regional strategy associating the semi-arid regions with the rest of the country. Efforts must therefore include the whole country and must be controlled. The present situation, in which spontaneous migrations take place in order to cope with drought, would thus be avoided. There has been no technical progress in abandoned areas and in reception areas there is a proliferation of suburban shantytowns. Migrations should not be accepted as the work of fate.

Initiatives taken by some Latin American Governments are worth mentioning in this regard. The idea is to reduce population pressure on arid regions by developing humid tropical areas; for example, development of the Amazon with people from northeastern Brazil; the Bolivian *yungas* with the help of the people of the arid altiplano; humid tropical zones of Mexico with the help of the people of the arid plateaus. Different problems may arise, but these examples demonstrate the need for integrated national land use policies and illustrate the desirable complementarity of different climatic zones.

Marked improvements in agricultural production are possible in semi-arid zones. For rainfed cultivation, efforts should be concentrated on the growing of food crops, which calls for close integration of agriculture and animal husbandry.

Types of integration of agriculture and animal husbandry

In most arid and semi-arid zones, agriculture and animal husbandry compete for the use of land. It is essential for development not only to take technical measures, but to define the respective areas of these two activities, and to ensure their harmonious coexistence. The association of agriculture and animal husbandry in the same areas and, if possible, in the same holdings, would be of considerable interest from an economic and technical (especially agronomic) point of view. This possibility appears, however, to be limited to two relatively marginal areas: irrigated areas and the wetter fringes of the arid zone.

In irrigated areas, it has been shown that the development of fodder crops depends on an increase in the prices of livestock produce. But thought is seldom given to using animal traction for irrigation, even though pumped irrigation is often limited by the cost. A form of integration which could be introduced in small holdings would therefore be to use animal traction for pumping the water for irrigation (and of course for the animals to drink), as well as for drawing the ploughs. This would be an inexpensive way of extending the area of land under irrigated farming along the banks of certain rivers.

At the edges of all semi-arid regions, animal husbandry can benefit from the by-products of agriculture and at the same time provide farmers with manure and traction power. There are historical reasons why it is difficult to associate these two activities (competition between herdsmen and sedentary populations), in addition to certain cultural features (the fact that Black Africa traditionally used neither the wheel nor animal traction), technical problems (difficulty of training animals which are moved from place to place) and sociological shortcomings (lack of institutions to co-ordinate animal husbandry and crop-farming). It is on the southern borders of the Sahara that all these difficulties are combined, a situation rendered all the more paradoxical by the fact that there are more cattle there than in other areas. *Acacia albida*, which is instrumental in the development of sedentary farming of both crops and livestock, is also found in this region.

This tree, the importance of which has already been stressed, is conducive to soil fertilization and continuous cropping, and also allows cattle to remain in one area, by providing them with nutritive aerial fodder in the dry season. It is only populations with an established agro-pastoral tradition which really understand the role this tree plays and which take steps to maintain and spread it, since in traditional systems, it is the cattle that are responsible for its spread, as the seeds germinate only after they have been eaten by ruminants. But modern man understands how to ensure its germination and plantation. A policy to promote simultaneously fixation and fertilization of cultivated lands, to maintain sedentary herds and use them as a source of organic matter, power, meat and milk, and the re-afforestation and management of the area might be based on *Acacia albida*; this is clear from analysis of the high-density patches in the Sahel-Sudanian zone. It is, however, impossible to determine precisely the potential range of this tree. It is not known what degree of aridity it can tolerate or what maximum groundwater depth. But it can be said that *Acacia albida* is very valuable for the development of both agriculture and animal husbandry in all regions where the minimum summer rainfall is 400 mm, and where the groundwater table is accessible in the dry season.

NON-ZONAL FEATURES: INDUSTRIALIZATION AND URBANIZATION

Industrial development

Some arid and semi-arid zones have energy and mineral resources, whereas others do not. The prospects for industrial development are thus very uneven. In addition, and whatever the amount of raw materials, consideration must be given to the geographic position of deserts and their fringes in relation to the vital centres of the countries concerned. Training facilities, and especially potential markets, are rarely found in the desert, except in a few isolated instances (the shores of the Persian Gulf and the coastal cities of Peru). That is why industrialization like that of the western United States, resulting in the creation of urban oases with a concentration of highly technical, very productive, light industries (Phoenix, Tucson) cannot be taken as a model by most countries.

Moreover, industrialization (i.e. a series of integrated activities rather than a simple mining operation) is not solely dependent on the existence of mineral deposits. Production costs and price structures must be carefully examined. Steelworks can scarcely be set up in the middle of the desert, far from markets, even though iron ore and energy may abound. The gas and oil of Algeria, for instance, are used in the urbanized coastal areas, far from their original extraction sites. Economic reasons, as well as the conditions specific to an arid environment, leave no alternative other than an urban location, particularly in the case of basic industries. Densely populated semi-arid environments, however, show greater possibilities, especially for light industries producing consumer goods.

In industrial nations, mining has often been a stimulant because it was integrated to an overall development process. Under present day conditions, the mining economy of arid and semi-arid countries (importance of foreign investments, exporting of hydrocarbons and basic ores) cannot hope to promote integrated development. Geared to foreign needs, these industries have few linkages with the rest of the economy. The co-ordination of economically complementary units depends more on the nature of the economic system and the logic upon which growth is founded, than on simple existence of production units. In other words, the construction of a steel work, or of any other heavy industrial unit, does not guarantee the creation of an industrial region.

Moreover, the foreseeable depletion of some resources should be considered. It has been estimated that currently exploited oil fields could be exhausted within 70 to 80 years; natural gas within 50 years. Overuse of minerals useful to rich countries will first affect non-industrial exporting countries; if the growth of consumption continues at present rates, known reserves of copper will be depleted in 40 years, those for lead in 15 years and zinc reserves in 18 years. The products derived from mineral resources cannot sustain long-term growth. In addition, they are vulnerable to technological evolution. Thus, national use of other resources must be contemplated. In this context, and excluding exceptional circumstances (Kuwait, United Arab Emirates), the desert can only be used as a service area, but it should benefit in return from investments for irrigation or the reorganization of animal husbandry.

However, a number of possibilities for the industrialization of countries of the arid and semi-arid world may be considered when industries are well chosen.

Semi-arid countries

In the densely populated semi-arid fringes where raw materials are scarce, a major opportunity lies in the establishment of agro-industrial complexes connected to improved, surplus-generating, food crop production and animal husbandry (flour mills, meat processing industries, milk and animal by-products). These industrial units could ease the demographic pressure on cultivated areas while locally holding back part of the manpower by providing jobs. This would reduce rural migration towards large urban centres, ill-adapted to receive it. These types of light industry, which never reach the gigantic proportions of certain basic industries, require few investments and yield a rapid capital turnover. Finally, progress in land transport, particularly by truck, as a consequence of the extension of the road networks and the improvement of dirt tracks, helps flexible location policies.

Although the proximity of energy is no longer essential to the establishment of a plant, the power supply remains a major obstacle. At the present stage of research, neither wind nor solar energy is widely used. The former requires windy sites, cannot be stored, and is very unreliable because of changing winds. The latter involves investments which at present are far too costly. However, the considerable insolation of arid and semi-arid zones could, in the long term, become a positive factor. Deserts and their fringes, because of their latitude and generally low nebulosity, benefit from long periods of sunshine which add up to 3,000 or 4,000 hours a year and receive between 300 and 650 calories/cm^2/day. Thus they lend themselves to studies aimed at the improvement of existing techniques, particularly photochemical and photoelectric conversion, the costs and outputs of which still remain prohibitive. (The cost of an installed photovoltaic cell is a million dollars per kW.) The only practical application today would be the production of power on a small scale to operate domestic equipment and possibly small industries. However, the power supply problem can be partially solved by developing the production of hydro-electric power in connection with irrigation programmes or by tapping potential hydrocarbon beds in nearby deserts.

Arid countries

Where desert occupies all or part of the national territory, the situation is more complex. Though many deserts possess considerable energy resources, the distribution of ore deposits is very uneven. The size and population of states are also factors to be taken into account. Thus the investment potential of Kuwait and Libya (small and large areas, but both sparsely populated) is distinctly higher than actual industrialization possibilities. Iran, on the other hand, with a high population and considerable mineral resources, seems to be in a better position.

For the arid oil-producing countries several suggestions can be made. In the case of countries with insufficient manpower and a small domestic market, industrial development might be directed towards sectors requiring heavy energy consumption, or towards petrochemical industries linked to oil transport (naval repair yards). This sort of undertaking is possible in countries of the Persian Gulf where power is abundant and desalination plants are already in operation. (Bahrein, Kuwait, Qatar).

The development of industry depends, however, on marketing possibilities. A strategy based on the construction of petrochemical complexes destined to export their production, may face a saturated world chemical market. Carrying out all those projects which have been put forward to date would result in this catastrophic situation rendering the capital equipment of these countries useless. For countries with a plentiful labour force, greater diversification could be considered in the context of an association between existing consumer goods industries, capital goods industries with high value added, and considerable linkages (e.g. transport, electrical or agricultural equipment) and intensive agriculture which, by present day standards, would not necessarily be considered profitable.

Relations with industrialized countries

Whether raw materials are located in the desert or on its periphery, it is essential to process resources in their country of origin. It is equally important to establish industrial production on a level that is not limited exclusively to national markets. Thus interregional and international exchanges are a fundamental condition of industrial development. Since 1945, most industrialization policies have been based on the concept of technology transfer, which has been set up as dogma in industrial development economics. Following this concept a "take off" point (Rostow's "take off") is established among the stages of development, which determines the volumes of investments.

However, the transfer of technology, which includes knowledge and knowhow, faces a number of obstacles. The scientific level of today's technology contributes to widening the gap between what industrialized countries have to offer and what poor countries are able to assimilate. In the early stages transfers were concerned mainly with patents and manufacturing licenses. This created a new kind of "technological dependence". At the second stage, transfers concerned the sale of turnkey factories, and today industrial companies furnish the market along with the factory.

Under these conditions there is a great risk of prestige projects, ill-adapted to the needs of the host countries. In addition, transfers generate new expenses, which are linked to multiple royalties and exceptional working costs resulting from production agreements (increasing imports of semi-finished products, high cost of intermediate consumption, need to dump exports, tax avoidance linked to the ownership of the technologies).

Moreover, low management capacity and the difficulties encountered in training skilled labour, show that the problems of industrial development are not only technical ones. The arid and semi-arid countries are not "new lands"; the heritage of civilization remains considerable and the possible use of oil profits for the necessary large investments is not the only element to be taken into consideration. The implementation of an industrial development policy should take account of this historical and social heritage.

Finally, the study of the problems of industrial development should not be limited to arid zone constraints. The mechanisms responsible for the present lack of industrialization are attributable not so much to the effects of heat or rainfall as they are to the nature of the relations between industrialized countries and the rest of the world.

Urban development

Several series of measures could help regulate urban expansion.

Regionalization

In industrial countries the presence of an urban network, or a hierarchy of cities maintaining functional relations, is often the consequence of economic development. These networks do not exist, however, in highly-populated semi-arid countries. The attraction of large cities is so strong that there is a real danger of an over-concentration of the population in one or two points of the territory.

Medium-size towns located in regions once occupied by populations now in the process of urbanization, play only a small role in regionalization for the moment. Often inhabited by absentee landlords, or by impoverished farmers, they are in most cases no more than temporary stops for migrants on their way to the capital. For this reason their development appears as a useful alternative to topheavy capital cities. This implies a geographic redistribution of functions and jobs in the service sector, as well as decision powers at a regional level. This regionalization policy could be based on the creation of jobs in industry in connection with agricultural development programmes. The position of new industries should be chosen, however, in such a way that their water supply does not compete with agriculture. In this context the modernization of the transport network, in conjunction with an effort to increase the number of motor vehicles, particularly heavy vehicles, could provide a large variety of choices by providing medium-size towns with a dynamic role in regional development: collecting of raw materials, meat and forage markets, various redistributions in connection with the capital, processing of agricultural products and related activities.

Thus, medium-size towns could act as a regional filter by absorbing part of the migration stream headed for the capital and by contributing to a more just geographic distribution of development.

Urban planning

Within urban centres, land ownership control must be established in order to prevent speculation on building lands and to insure that this land is not developed at the expense of rich urgently needed farmland. This control system should regulate the use of urban land (by nationalization, municipalization, or any other means at the service of the community). In this respect the preparation of master plans, projects for land use and zoning, and other forms of planning, are a basic requirement for any effort to put an end to the anarchy of urban growth.

This anarchy is reflected today in mushrooming shantytowns and a deterioration in urban living conditions. To stop this downward trend states must first of all equip themselves with effective statutory and legal weapons which could become instrumental in an urban strategy serving all city dwellers.

In this context, it seems possible to implement a housing policy (renovation or development) geared to the needs of low income populations. Urban centres containing millions cannot be transformed totally nor rapidly: thus transitional solutions should be considered. This does not mean defending the *status quo* in the shantytowns any more than it means proposing rapid transformation requiring investments which most countries cannot afford. It is obvious that a complete renovation of antiquated urban sectors and total control of urban growth are conceivable only in connection with overall economic development. Similarly, the improvement of housing entails an increase in family income, which in turn implies economic progress at a national level, as well as a just redistribution of income.

The various urban experiments conducted in Latin America, Asia or Africa may be of benefit to the cities of the arid and semi-arid world. The failure of housing policies favouring prefabricated buildings (not among civil servants, but among the poor) should be studied carefully. The insufficiency of the occupiers' income, due to the scarceness of resources, has in most cases resulted in rapid deterioration of these buildings from lack of maintenance. Moreover, this type of housing is often poorly suited to the heirs of agricultural societies who maintain rural practices within urban centres (gardening and rearing a few animals).

However, some improvements have been achieved when administrations have taken charge of infrastructure works and communal facilities (road systems, water, electricity), while the inhabitants construct their own dwellings. With this in mind a credit policy could facilitate the purchase of building materials and promote, for example, the establishment of housing co-operatives, which to credit organizations would appear more solvent than individuals. Thus the search for a temporary equilibrium between housing and resources could help avoid the type of waste characteristic of programmes which are often ambitious but ill-adapted to the needs of the poor.

Moreover, a global housing policy should adopt an architectural style. It is therefore incumbent on national architects to take responsibility for the conception and implementation of housing programmes. In this context, the use of local materials and the adaptation of traditional heat reducing techniques could be of great use.

The application of new techniques using solar energy seems desirable. In some countries (Australia, Israel, United States) domestic needs such as air conditioning, refrigeration and conservation of food, have already been covered. Small-scale use of solar energy would solve serious problems involving communal facilities, as well as improve the technical knowledge of the population.

Finally, urban development is inconceivable without a water supply system which takes into account the limited resources. For this reason, in addition to the research concerning brackish water, systematic recycling of water must be considered. A water supply policy should also include a sewage disposal and refuse treatment programme (WHO 1974).

IV. CONCLUSION: LIMITS AND POSSIBILITIES

This review of obstacles to development in arid and semi-arid lands touches a wide variety of situations and problems to which scientific and technological research is directed. As a result of the work completed to date the dimensions of several of the distinctive problems of arid and semi-arid zones are now understood more clearly than when the first international attack on them was launched 25 years ago under the auspices of Unesco. The UN Water Conference, and the Conference on Desertification (1977) will enable a review of our state of knowledge in these fields. In this respect it is important to underline the particular co-ordinating role which UNEP has played in this field since 1973. Human capacity to deal with those problems has increased, and scientific findings have been applied in various ways. In some areas the quality of life has been enhanced. In others the situation remains unchanged. In still others well-intentioned efforts at development have degraded and compromised both human welfare and the resource base.

It has been shown that the development of stable agriculture - whether by grazing or by irrigation - is far more complex than had been thought by enthusiasts who had expected single technical solutions, such as wells, desalination, artificial rainfall, new seeds, and tree planting, to transform arid lands. Some problems, such as the maintenance of plant cover in steppe lands of high population density have proved intractable so long as basic changes are not made in social organization or in settlement patterns. The difficulties are now recognized.

During the quarter century since 1950, the technical capacity to make use of arid and semi-arid zone resources changed significantly in some directions and not in others. New hypotheses on climatic variability are shaking development plans based on conditions of the preceding 10-20 years, conditions which were expected to continue in the future. Understanding of the ecological relationships between soil, water and plants has improved. Progress has been made in the mapping and inventory of resources and in the new techniques of remote sensing, especially by satellite. Simulation modelling of groundwater and of surface water has become a useful tool; improvements have been made in methods of increasing water supply by rainfed farming, saline water irrigation, recycling of waste water, and well drilling. Measures to conserve water by reducing losses, innovations in irrigation methods, and controlled-environment cultivation of crops have become usable in a few areas.

Major attention to plant breeding has led to new strains of high-yield drought-resistant crops. Effective methods have been found to forecast and cope with movements of the desert locust. Some improvements have been made in dealing with animal disease and in curbing river blindness. Progress has also been made in public health and the need is recognized for well-designed and detailed studies on factors which affect populations obliged to live in arid conditions.

The solutions suggested in this report are possible if they are supported by research programmes requiring less money and time. They involve particularly the creation of new varieties and the development of techniques to increase agricultural and pastoral yields conditions where water is lacking and there is a risk of salinity. They also involve the various aspects of water science and technology, as well as urban technology and tropical medicine in a dry or irrigated environment. On the other hand, other research has not justified the hope it had raised. Artificial rainfall by cloud seeding still raises many problems in many regions and the techniques are implemented without a complete knowledge of the physical processes involved. Desalination of sea water and the conversion of solar energy can be satisfactorily undertaken only in a few regions. Innovations in the conception and construction of urban housing remain limited. The transfer of new technology and scientific discoveries for environmental management and the organization of human establishments have been very slow.

TRANSFER OF TECHNOLOGY

Although the transfer of technology is essential to modern development of arid and semi-arid regions, the present situation is far from satisfactory for several reasons :
- The environments involved are not all the same. Thus recent the achievements and experiments in Australia or the United States concerning the rationalization of animal husbandry, have been in regions not subject to the same climatic uncertainty as pastoral regions in the old world. In the latter, pastoralists have often been driven out by cereal farmers towards more marginal zones, whereas in countries of the new world animal husbandry is often found in semi-arid conditions, which do not exclude cereal cultivation.
- The proposed technology is increasingly complex. The results of the green revolution have demonstrated the limitations of a hasty application of such advanced techniques in a social environment poorly prepared to receive them. Though some rich farmers have profited from them, the majority have not had the financial means to follow their example.
- Among developing countries with semi-arid regions, it is important to distinguish those which already have a solid technical and scientific infrastructure (Brazil, India, Mexico) and are better able to assimilate new techniques.
- In many cases of technology transfer, the transfer involves only one stage of the technology, thus creating a dependence which may in fact hinder true development. Ignoring cost, an important requirement is that the transferred technology be adapted to the needs and situation of the country which receives it (Marques dos Santos 1975). The ideal would be technology exchange, not transfer. For the time being, and for most of the countries concerned, the advantages of intermediate techniques deserve mention. These techniques, according to Myrdal (1974), are not those which are obsolete in industrialized countries, but rather those which while applying the latest results of scientific research are best suited to maximum use of the labour force.

The problem of adaptation to socio-cultural environments in the host countries, as well as that of harmonization with local technical experiments, remains. The creation of a context favourable to the introduction of new techniques is a long process requiring appropriate education programmes. With this in mind adult education programmes for the working population, linked to development programmes, seem better suited to the desired transfers than the present school system.

RESEARCH NEEDS

The amount of research is still notably insufficient. Thus, the lowering of the cost of a kW of solar energy to a fiftieth of its present value is no more difficult than controlled nuclear fusion, provided comparable funds and sufficient time are allocated to it. Such an achievement would substantially modify the economy of arid zones.

The main gaps in knowledge, and its application, could either be filled in the short or medium term, or only in the long term. In any case true development is not linked to research in a particular sector, but rather to an overall organization and mastery of the environment.

In the short and medium term, better use of scarce water resources, which are irregular in time and in space, depends on a better understanding of climate. More observations will allow maps to be revised and more accurate measurement, at least in some areas, of the little known water balances in the main types of soil. In addition, groundwater and its

Conclusion : limits and possibilities

recharge, the quality of the water and the effect of more or less intense use on its levels, should be the subject of precise hydrogeological studies. In irrigated areas, effective control of salination and water-logging needs a more thorough application of existing techniques. For non-irrigated crops, the selection of varieties suitable for human or animal consumption, with a short cycle, low water requirements and good adaptation to a given rainfall regime, seem possible in the near future. In this respect most attention should be paid to varieties able to be diffused in the context of traditional agriculture, rather than to "miracle" varieties, which are difficult to popularize outside experimental stations.

The fight against desertification obviously implies a better understanding of ecosystems, but above all systematic application of well-established protection and biological regeneration techniques.

Particular attention should be given to health conditions in irrigated areas. As numerous examples have shown, the establishment of effective water control is not incompatible with public health needs. Similarly, in cities suitable construction techniques can contribute to the improvement of health and living conditions.

As social obstacles to changing life styles are at least as important as technical ones, more systematic use of the various social sciences should be made in the future. The preparation of suitable education programmes must be based on this knowledge. In this context studies conducted by local research workers are likely to be more in touch with local realities, particularly psycho-social realities.

In the long term, considerable progress will certainly be made as soon as techniques for desalination of brackish, and then salt water, can provide an adequate water supply at reasonable prices. The mastery of solar energy would undoubtedly bring about other important changes, particularly in its potential applications to water management and for the desalination of brackish and eventually salt water. However, even when tangible results have been obtained, which largely depend on the investments devoted to research, all these technical advances will still have to be integrated to an overall strategy of land management.

The existence side by side of different land use systems implies not only their juxtaposition, but also numerous relationships of competition at the level of overall spatial organization. If nothing is changed in the future, the consequences of exceptional drought periods will be even more catastrophic, not as a result of increased aridity of the climate, but simply because of higher population densities, which are likely to increase conflicts between different types of land use, and to drive farmers towards less favourable areas and pastoralists towards marginal grazing lands. Thus any development strategy should be based on coherent overall planning of land and water use, the two being closely related in semi-arid zones. In this respect, precise studies could be conducted on:

- the rational use of water and fair distribution among the main consumers (agriculture, pastoralism, industry, tourism, towns, etc.) and at the same time constant supervision of the quality and quantity of irrigation water, and recycling of waste water;
- the fixing of climatic limits (with due consideration to soil conditions) to the spread of rainfed cultivation, especially mechanized cultivation, which seems to be more detrimental to the environment;
- the demarcation of rangelands in order to organize rotational use and at the same time links between pastoralism and irrigated areas;
- the protection of wooded areas, particularly around the main towns, but also in all inhabited zones, in order to avoid the depletion of forests. Given the advanced degradation of some areas (for example, around Ouagadougou or Niamey, or the main cities of the Middle East) protection will not suffice and specific regeneration measures will have to be taken; maintenance of forests is also highly beneficial to agriculture and pastoralism; similarly creation of National Parks (e.g. the Azraq Park in Jordan) may contribute to the development of tourism (ALECSO-UNEP 1975);
- the regulation of urban growth.

It is obvious that the distribution of available space (and available water) calls for political choices and political judgements. Global development plans which take into account the links between different fields of activity (for example, the development plan for Lake Nasser, which allows for simultaneous development of fishing, agriculture, transport and tourism, in conjunction with protection of public health) are preferable to sectorial actions (FAO 1975). Such overall plans for land and water management must be accepted by the various populations involved. This calls for a suitable institutional framework representing the interests and aspirations of the people, in which the latter also share responsibility for the proper application of decisions taken as a result of consultation.

Conclusion : limits and possibilities

This last point introduces the problem of how to measure development. Profitability cannot be the only reference point, since it does not have the same meaning for a farmer using advanced techniques on his own farm, for a fellah working with a hoe on the land of an absentee landlord, or for a semi-nomad, who in addition to "high-risk cultivation" is practicing animal husbandry with highly variable results from one year to the next. The construction of an enourmous dam, or the establishment of a large factory or luxury resort area should not be taken in themselves as signs of development; success is, rather, a measure of the improvement in the local population's living conditions and of the distribution of income from economic transformations.

It has become evident that while there are benefits to be realized from further research, the most urgent need is for better integrated approaches to the planning and management of development programmes. The record shows that those efforts, including those which lead to initiative by individual farmers or pastoralists, will not be effective if oriented wholly around purely technical solutions. Nor can it be assumed that the technical and scientific research can be translated readily into field applications. Most efforts have suffered from concentration on narrow sectors of study to the detriment of integrated approaches that take in the whole complex of resources and human needs of a region, and that look to practical use of the findings.

BIBLIOGRAPHY

*ACADEMIE DES SCIENCES D'OUTRE MER. 1975. La sécheresse en zone sahélienne. Causes, conséquences, étude des mesures à prendre. *NED*, No. 4216-4217. La Documentation française, Paris.

*ALECSO-UNEP. 1975. *Re-greening of Arab Deserts*. ALECSO, Cairo.

AMIRAN, D.H.K.; WILSON, A.W. (Eds). 1973. *Coastal deserts : their natural and human environments*. University of Arizona Press, Tucson.

*ANDREAE, B. 1974. Die Farmwirtschaft an den Agronomischen Trockengrenzen. *Erdkundliches Wissen*, 38, No. VIII.

*BALBO, J.L. 1975. Sahel : mettre la culture irriguée entre les mains des paysans. *Actuel développement*, 6, p. 38-45.

*BATISSE, M. 1969. Problems facing arid-land nations. In : *Arid lands in perspective*. W.G. McGinnies and B.J. Goldman, p. 1-12. American Association for the Advancement of Science, Washington and University of Arizona Press, Tucson.

*BERNUS, E.; BERNUS, S. 1972. *Du sel et des dattes : introduction à l'étude de la Communauté d'In Gall et de Tegidda-n-tesemt*. Etudes nigériennes, No. 31. Centre nigérien de recherches en sciences humaines, Niamey.

BOUDET, G. 1972. Désertification de l'Afrique tropicale sèche. *Adansonia*, série 2, 12 (4), p. 504-524.

BROWN, A.W.A.; DEOM, J.O. 1973. Summary: health aspects of man-made lakes. In: *Man-made lakes: their problems and environmental effects*. Geophysical Monograph, 17. W.C. Ackerman, G.F. White and E.B. Worthington (Eds), p. 755-764. American Geophysical Union, Washington.

CLARKE, J.I.; FISHER, W.B. 1972. *Populations of the Middle East and North Africa. A geographical approach*. University of London Press, London.

COOKE, R.U.; WARREN, A. 1973. *Geomorphology in deserts*. Batsford, London.

DASMAN, R.F.; MILTON, J.P.; FREEMAN, P.H. 1973. *Ecological principles for economic development*. John Wiley and Sons Ltd, London, New York, Sydney, Toronto.

DELWAULLE, J.C., 1973. Désertification de l'Afrique du Sud du Sahara. *Bois et forêts des tropiques*, 149, p. 3-20.

*DIRECTION GENERALE DE LA RECHERCHE SCIENTIFIQUE ET TECHNIQUE (DGRST); SOCIETE POUR L'ETUDE DU DEVELOPPEMENT ECONOMIQUE ET SOCIAL (SEDES). 1976. *Eléments de statistiques pour une analyse du développement socio-économique dans les six pays du Sahel*. SEDES, Paris.

DREGNE, H.E. (Ed.). 1970. *Arid lands in transition*. American Association for the Advancement of Science, Washington.

ERIKSSON, E.; GUSTAFSSON, Y.; NILSSON, K. (Eds). 1968. *Groundwater problems*. Proceedings of the international symposium held in Stockholm in October 1966. Wenner-Gren Center International Symposium Series, volume 11, Pergamon Press, London.

* *References cited in this Technical Note.*

Bibliography

ETIENNE, G. 1967. *Progrès agricoles et maîtrise de l'eau: le cas du Pakistan.* Paris.

EVENARI, M. 1971. *The Negev. The challenge of a desert.* Harvard University Press, Cambridge.

FAO. 1972. La salinité. *Bulletin d'irrigation et de drainage,* 7.

FAO. 1974a. *Aménagement écologique des terrains de parcours de l'Afrique et du Moyen Orient.* Rapport d'une consultation d'experts. FAO, Rome.

FAO. 1974b. *Amélioration de la production dans des zones de basse pluviométrie.* COAG. FAO, Rome.

*FAO. 1975. *Project findings and recommendations.* Lake Nasser Development Centre, Aswan, Egypt.

FAO-UNESCO. 1973. *Irrigation, drainage and salinity. An international source book.* Hutchinson/FAO/Unesco, London.

FLORET, C.; LE FLOC'H, E. 1973. *Production, sensibilité et évolution de la végétation et du milieu en Tunisie présaharienne.* Document C.E.P.E. No. 71. Centre d'Etudes Phytosociologiques et Ecologiques Louis Emberger et Institut National de la Recherche Agronomique de Tunisie, Montpellier and Tunis.

GALLAIS, J. 1972. Essai sur la situation actuelle des relations entre pasteurs et paysans dans le Sahel ouest-africain. In : *Etudes de géographie tropicale offertes à P. Gourou,* p. 301-313. Mouton, Paris.

GOOR, A.Y.; BARNEY, C.W. 1976. *Forest tree planting in arid zones.* Ronald Press Co., New York.

GROVE, A.F. 1973. Desertification in the African environment. In : *Drought in Africa. Report of the 1973 Symposium.* D. Dalby and R.J. Harrison (Eds), p. 33-45. University of London, London.

HEINDL, L.A. (Ed.). 1975. *Hidden waters in arid lands.* Report of a workshop on groundwater research needs in arid and semi-arid zones held in Paris, France, 25 November 1974. International Development Research Centre, Ottawa.

HILLS, E.S. (Ed.). 1966. *Arid lands. A geographical appraisal.* Methuen and Co. and Unesco, London and Paris.

HODGE, C.; DUISBERG, P.C. (Eds). 1963. *Aridity and man. The challenge of the arid lands in the United States.* Publication No. 74. American Association for the Advancement of Science, Washington.

IRONS, W.; NEVILLE, D. 1972. *Perspectives on nomadism.* Leiden.

KASSAS, M. 1970. Desertification versus potential recovery in circum-Saharan territories. In: *Arid lands in transition.* H.E. Dregne (Ed.), p. 123-142. American Association for the Advancement of Science, Washington.

LE HOUEROU, H.N. 1972. An assessment of the primary and secondary production of the arid grazing land ecosystems of North Africa. In : *Proceedings of the International Symposium on the Ecophysical foundation of ecosystems productivity in arid zones,* p. 168-172. USSR Academy of Sciences, Leningrad.

LE HOUEROU, H.N. 1973. Ecologie, démographie et production agricole dans les pays méditerranéens du Tiers-Monde. *Options Médit.,* 17, p. 53-61.

LE HOUEROU, H.N.; COSTE, C.H. 1976. *Relationships between rangeland production and average annual rainfall.* Volume I: Mediterranean Basin. Volume II: Sahelian and Sudanian zones of Africa.

LEROUX, M. 1973. *La saison des pluies 1973 au Sénégal.* ASECNA Publication No. 32, Dakar.

LESHNIK, L.S.; SONTHEIMER, G.D. 1975. *Pastoralists and nomads in South Asia.* Wiesbaden.

LONG, G. 1974. *Diagnostic écologique et aménagement du territoire.* Tome 1. Masson et Cie, Paris.

LONG, G. 1975a. *Diagnostic écologique et aménagement du territoire.* Tome 2. Masson et Cie, Paris.

LONG, G. 1975b. Pour une stratégie de la recherche, dans le cadre du Projet 3 du MAB, appliquée aux zones arides au nord du Sahara. *Options Médit.,* 26, p. 39-50.

*MARQUES DOS SANTOS, A. 1975. Les mécanismes actuels du transfert de technologie sont-ils favorables aux P.V.D.? In: *Actuel développement,* 9, p. 26-34.

McGINNIES, W.G.; GOLDMAN, B.J. 1969. *Arid lands in perspective.* American Association for the Advancement of Science, Washington and University of Arizona Press, Tucson.

McGINNIES, W.G.; GOLDMAN, B.J.; PAYLORE, P. (Eds). 1968. *Deserts of the world: an appraisal of research into their physical and biological environments.* University of Arizona Press, Tucson.

McGINNIES, W.G.; GOLDMAN, B.J.; PAYLORE, P. (Eds). 1971. *Food, fiber and the arid lands.* University of Arizona Press, Tucson.

MEIGS, P. 1966. *Geography of coastal deserts.* Arid zone research XXVIII. Unesco, Paris.

MENSCHING, H.; WIRTH, E. 1973. *Nordafrika. Vorderasien.* Fischer Länderkunde Collection, Volume 4. Fischer-Verlag, Frankfurt.

MONOD, T. 1973. *Les déserts.* Horizons de France.

MONOD, T. (Ed.). 1975. *Pastoralism in tropical Africa.* Oxford University Press, London.

*MYRDAL, G. 1974. The transfer of technology to underdeveloped countries. *Scientific American,* 3, p.172-182.

NATIONAL ACADEMY OF SCIENCES. 1974. *More water for arid lands. Promising technologies and research opportunities.* Report of an ad hoc panel of the Advisory Committee on Technology Innovation, Board on Science and Technology for International Development, Commission on International Relations, Washington.

NIR, D. 1974. *The semi-arid world.* Longmans, New York.

*NOIN, D. 1971. *La population rurale du Maroc: étude géographique.* Presses Universitaires de France, Paris.

NOY-MEIR, I. 1975. *Primary and secondary production in sedentary and nomadic grazing systems in the semi-arid region: analysis and modelling.* Final Research Report on Project 7/E-3 submitted to the Ford Foundation.

PELISSIER, P. 1966. *Les paysans du Sénégal; les civilisations agraires du Cayor à la Casamance.* St Yrieix.

*PETTERSSEN, S. 1941. *Introduction to meteorology.*

PLANHOL, X. de; ROGNON, P. 1970. *Les zones tropicales arides et subtropicales.* A. Colin, Paris.

RAPP, A. 1974. *A review of desertization in Africa: water, vegetation and man.* SIES Report No. 1. Secretariat for International Ecology, Stockholm.

*SASSON, A. 1970. Terres arides dans un monde en mutation. In : *Comptes-rendus des séances mensuelles de la Société des Sciences naturelles et physiques du Maroc,* p. 35-66.

SCHIFFERS, H. 1971-1973. Die Sahara und ihre Randgebiete. *Afrikastudien,* 60-62.

*SEBILLOTTE, M. 1973. Les cultures de céréales en sec dans le Maroc oriental. *Revue de géographie du Maroc,* 23-24, p. 51-77.

SHERBROOKE, W.C.; PAYLORE, P. 1973. *World desertification: cause and effect. A literature review and annotated bibliography.* Arid Lands Resource Information Paper 3. University of Arizona, Tucson.

Bibliography

UNESCO. 1957. *Guide book to research data for arid zone development.* Arid Zone Research IX. Unesco, Paris.

UNESCO. 1961. *A history of land use in arid regions.* Arid Zone Research XVII. Unesco, Paris.

UNESCO. 1962. *The problems of the arid zone. Proceedings of the Paris Symposium.* Arid Zone Research XVIII. Unesco, Paris.

UNESCO. 1963a. *Environmental physiology and psychology in arid conditions. Reviews of research.* Arid Zone Research XXII. Unesco, Paris.

UNESCO. 1963b. *Nomades et nomadisme au Sahara.* Recherches sur la zone aride XIX. Unesco, Paris.

UNESCO. 1963c. *Changes of climate. Proceedings of the Rome Symposium organized by Unesco and the World Meteorological Organization/Les changements de climat. Actes du colloque de Rome organisé par l'Unesco et l'Organisation météorologique mondiale.* Arid Zone Research XX/Recherches sur la zone aride XX. Unesco, Paris.

UNESCO. 1964. *Environmental physiology and psychology in arid conditions. Proceedings of the Lucknow Symposium/Physiologie et psychologie en milieu aride. Actes du colloque de Lucknow.* Arid Zone Research XXIV/Recherches sur la zone aride XXIV. Unesco, Paris.

UNESCO. 1965. *Methodology of plant eco-physiology. Proceedings of the Montpellier Symposium/Méthodologie de l'éco-physiologie végétale. Actes du colloque de Montpellier.* Arid Zone Research XXV/Recherches sur la zone aride XXV. Unesco, Paris.

UNESCO. 1975. *The Sahel: ecological approaches to land use.* MAB Technical Notes 1. Unesco, Paris.

*UNESCO-FAO (In publication since 1971). *Soil map of the world.* Unesco, Paris and FAO, Rome.

VAN KEULEN, H. 1975. *Simulation of water use and herbage growth in arid regions.* Centre for Agricultural Publishing and Documentation, Wageningen.

WHITE, G.F. (Ed.). 1956. *The future of arid lands.* Publication No. 43. American Association for the Advancement of Science, Washington.

*WHO. 1974. *Evacuation des eaux usées des collectivités.* Rapport technique No. 541. WHO, Genève.

*WORTHINGTON, E.B.(Ed.) 1977. *Arid land irrigation in developing countries. Environmental problems and effects.* Based on the international symposium, 16-21 February 1976. Alexandria, Egypt, Pergamon Press, London.

YARON, B.; DANFORS, E.; VAADIA, Y. (Eds). 1973. *Arid zone irrigation.* Ecological Studies 5. Springer-Verlag, Berlin, Heidelberg, New York.

Unesco publications: national distributors (Abridged list)

Argentina	EDILYR, Belgrano 2786-88, BUENOS AIRES.
Australia	*Publications:* Educational Supplies Pty. Ltd., Box 33, Post Office, BROOKVALE 2100, N.S.W. *Periodicals:* Dominie Pty. Ltd., Box 33, Post Office, BROOKVALE 2100, N.S.W. *Sub-agent:* United Nations Association of Australia (Victorian Division), 5th Floor, 134-136 Flinders Street, MELBOURNE 3000.
Austria	Dr. Franz Hain, Verlags -und Kommissionsbuchhandlung, Industriehof Stadlau, Dr. Otto-Neurath-Gasse 5, 120 WIEN.
Brazil	Fundação Getúlio Vargas, Serviço de Publicações, caixa postal 21120, Praia de Botafogo 188, RIO DE JANEIRO (GB).
Burma	Trade Corporation no. (9), 550-552 Merchant Street, RANGOON.
Canada	Renouf Publishing Company Ltd., 2182 St. Catherine Street West, MONTREAL, Que. H3H 1M7.
Chile	Bibliocentro Ltda., casilla 13731, Huérfanos 1160 of. 213, SANTIAGO (21).
Cuba	Instituto Cubano del Libro, Centro de Importación, Obispo 461, LA HABANA.
Czechoslovakia	SNTL, Spalena 51, PRAHA I (Permanent display); Zahranicni literatura, 11 Soukenicka, PRAHA I. *For Slovakia only:* Alfa Verlag, Publishers, Hurbanovo nam. 6, 893 31 BRATISLAVA.
Denmark	Ejnar Munksgaard Ltd., 6 Nørregade, 1165 KØBENHAVN K.
Ecuador	Casa de la Cultura Ecuatoriana, Núcleo del Guayas, Pedro Moncayo y 9 de Octubre, casilla de correo 3542, GUAYAQUIL. RAID de Publicaciones, García 420 y 6 de Diciembre, Casilla 3853, QUITO.
Ethiopia	Ethiopian National Agency for Unesco, P.O. Box 2996, ADDIS ABABA.
Finland	Akateeminen Kirjakauppa, Keskuskatu 1, 00100 HELSINKI 10.
France	Librairie de l'Unesco, 7, place de Fontenoy, 75700 PARIS, C.C.P. Paris 12598-48.
German Democratic Republic	International bookshops or Buchhaus Leipzig, Postfach 140, 701 LEIPZIG.
Germany (Fed. Rep.)	Verlag Dokumentation, Pössenbachestrasse 2, 8000 MÜNCHEN 71 (Prinz Ludwigshöhe). *'Unesco Courier' (German edition only):* Colmanstrasse 22, 5300 BONN. *For scientific maps only:* GEO Center, Postfach 800830, 7000 STUTTGART 80.
Ghana	Presbyterian Bookshop Depot Ltd., P.O. Box 195, ACCRA; Ghana Book Suppliers Ltd., P.O. Box 7869, ACCRA; The University Bookshop of Ghana, ACCRA; The University Bookshop, CAPE COAST; The University Bookshop, P.O. Box 1, LEGON.
Greece	Large Athens bookshops (Eleftheroudakis, Kauffman, etc.).
Hong Kong	Federal Publications Division, Far East Publications Ltd, 5A Evergreen Industrial Mansion, Wong Chuk Hang Road, ABERDEEN; Swindon Book Co., 13-15 Lock Road, KOWLOON.
Hungary	Akadémiai Könyvesbolt, Váci u. 22, BUDAPEST V; A.K. V. Könyvtárosok Boltjá, Népköztársaság utja 16, BUDAPEST VI.
Iceland	Snæbjörn Jónsson & Co. H.F., Hafnarstraeti 9, REYKJAVIK.
India	Orient Longman Ltd, Kamani Marg, Ballard Estate, BOMBAY 400 038; 17 Chittaranjan Avenue, CALCUTTA 13; 36a Anna Salai, Mount Road, MADRAS 2; B-3/7 Asaf Ali Road, NEW DELHI 1; 80/1 Mahatma-Gandhi Road, BANGALORE-560001; 3-5-820 Hyderguda, HYDERABAD-500001. *Sub-depots:* Oxford Book & Stationery Co., 17 Park Street, CALCUTTA 700016, *and* Scindia House, NEW DELHI 110001; Publications Section, Ministry of Education and Social Welfare, 511 C-Wing, Shastri Bhaven, NEW DELHI 110001.
Indonesia	Bhratara Publishers and Booksellers, 29 Jl. Oto Iskandardinata III, JAKARTA. Gramedia Bookshop, Jl. Gadjah Mada 109, JAKARTA. Indira P.T., 37 J1. Dr. Sam Ratulangi, JAKARTA PUSAT.
Iran	Commission nationale iranienne pour l'Unesco, avenue Iranchahr Chomali n° 300, B.P. 1533, TÉHÉRAN.
Ireland	The Educational Company of Ireland Ltd, Ballymount Road, Walkinstown, DUBLIN 12.
Israel	Emanuel Brown (formerly Blumstein's Bookstores), 35 Allenby Road *and* 48 Nachlat Benjamin Street, TEL AVIV; 9 Shlomzion Hamalka Street, JERUSALEM.
Jamaica	Sangster's Book Stores Ltd., P.O. Box 366, 101 Water Lane, KINGSTON.
Japan	Eastern Book Service Inc., C.P.O. Box 1728, TOKYO 100 92.
Kenya	East African Publishing House, P.O. Box 30571, NAIROBI.
Korea (Republic of)	Korean National Commission for Unesco, P.O. Box Central 64, SEOUL.
Libyan Arab Republic	Agency for Development of Publication and Distribution. P.O. Box 34-35, TRIPOLI.
Madagascar	Commission nationale de la République démocratique de Madagascar pour l'Unesco, B.P. 331, TANANARIVE.
Malaysia	Federal Publications Sdn Bhd., Balai Berita, 31 Jalan Riong, KUALA LUMPUR.
Malta	Sapienzas, 26 Republic Street, VALLETA.
Mexico	*For publications only:* CILA (Centro Interamericano de Libros Académicos), Sullivan 31 bis, MEXICO 4 D.F. *For publications and periodicals:* SABSA, Servicio a Bibliotecas, S.A., Insurgentes Sur n.° 1032-401, MÉXICO 12 D.F.
Netherlands	N. V. Martinus Nijhoff, Lange Voorhout 9, 's-GRAVENHAGE; Systemen Keesing, Ruysdaelstraat 71-75, AMSTERDAM 1007.
Netherlands Antilles	G. C. T. Van Dorp & Co. (Ned. Ant.) N. V., WILLEMSTAD (Curaçao, N.A.).
New Zealand	Government Printing Office, Government Bookshops: Rutland Street, P.O. Box 5344, AUCKLAND; 130 Oxford Terrace, P.O. Box 1721, CHRISTCHURCH; Alma Street, P.O. Box 857, HAMILTON; Princes Street, P.O. Box 1104. DUNEDIN; Mulgrave Street, Private Bag, WELLINGTON.
Nigeria	The University Bookshop, IFE; The University Bookshop, Ibadan, P.O. Box 286, IBADAN; The University Bookshop, NSUKKA; The University Bookshop, LAGOS; The Ahmadu Bello University Bookshop, ZARIA.
Norway	*All publications:* Johan Grundt Tanum, Karl Johans gate 41-43, OSLO 1. *For 'The Courier':* A/S Narvesens Litteraturtjeneste, Box 6125, OSLO 6.
Pakistan	Mirza Book Agency, 65 Shahrah Quaid-e-azam, P.O. Box 729, LAHORE 3.
Philippines	The Modern Book Co., 926 Rizal Avenue, P.O. Box 632, MANILA D-404.
Southern Rhodesia	Textbook Sales (PVT) Ltd., 67 Union Avenue, SALISBURY.
Singapore	Federal Publications (S) Pte Ltd., No. 1 New Industrial Road, off Upper Paya Lebar Road, SINGAPORE 19.
South Africa	Van Schaik's Bookstore (Pty.) Ltd., Libri Building, Church Street, P.O. Box 724, PRETORIA.
Spain	Ediciones Liber, Apartado 17, ONDÁRROA (Vizcaya); DONAIRE, Ronda de Outeiro, 20, Apartado de Correos, 341, LA CORUÑA; Librería Al-Andalus, Roldana, 1 y 3, SEVILLA 4; Mundi-Prensa Libros S.A., Castelló, 37, MADRID-1; LITEXSA; Librería Técnica Extranjera, Tuset, 8-10 (Edificio Monitor), BARCELONA.
Sudan	Al Bashir Bookshop, P.O. Box 1118, KHARTOUM.
Sweden	*All publications:* A/B C.E. Fritzes Kungl. Hovbokhandel, Fredsgatan 2, Box 16356, 103 27 STOCKHOLM 16. *For 'The Courier':* Svenska FN-Förbundet, Skolgränd 2, Box 150 50, S-104 65 STOCKHOLM.
Switzerland	Europa Verlag, Rämistrasse 5, 8024 ZÜRICH; Librairie Payot, 6. rue Grenus, 1211 GENÈVE 11.
Thailand	Nibondh and Co. Ltd., 40-52 Charoen Krung Road, Siyaeg Phaya Sri, P.O. Box 402, BANGKOK; Suksapan Panit, Mansion, 9, Rajdamnern Avenue, BANGKOK; Suksit Sim Company, 1715 Rama IV Road, BANGKOK.
Uganda	Uganda Bookshop, P.O. Box 145, KAMPALA.
U.S.S.R.	Mezhdunarodnaja Kniga, MOSKVA G-200.
United Kingdom	H.M. Stationery Office, P.O. Box 569, LONDON SE1 9NH; Government bookshops: London, Belfast, Birmingham, Bristol, Cardiff, Edinburgh, Manchester.
United Republic of Tanzania	Dar es Salaam Bookshop, P.O. Box 9030, DAR ES SALAAM.
United States	Unipub, Box 433, Murray Hill Station, NEW YORK, NY 10016.

A complete list of distributors is available from the Office of Publications, Unesco

Based on the work of a United Nations interagency task force, this Technical Note deals with the main physical, biological, sociological and institutional features associated with the management of arid and semi-arid lands. Specific attention is given to animal husbandry, agriculture, and urban, industrial and tourist development. For each of these subjects, obstacles to development, proposals for action, research needs and mechanisms for the transfer of technology are outlined.

This Technical Note should be of interest to those concerned with natural resources research and with the development of arid and semi-arid lands, as well as to administrators and students interested in these areas.

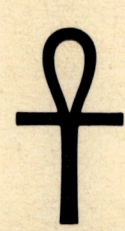

A stylized "ankh", the ancient Egyptian sign for life, has been incorporated into the symbol of the Programme on Man and the Biosphere (MAB).

ISBN 92-3-101484-6